Bernard L. Madison, University of Arkansas
Stuart Boersma, Central Washington University
Caren L. Diefenderfer, Hollins University
Shannon W. Dingman, University of Arkansas

Case Studies for Quantitative Reasoning
A Casebook of Media Articles

Third Edition

Supported by the National Science Foundation
Grant # DUE-0715039

Pearson Learning Solutions, 501 Boylston Street, Suite 900, Boston, MA 02116
A Pearson Education Company
www.pearsoned.com

Printed in the United States of America

5 6 7 8 9 10 VOCR 16 15 14

000200010271269018

BW/CM

ISBN 10: 1-256-51287-7
ISBN 13: 978-1-256-51287-5

Table of Contents

Introduction—Case Studies for Quantitative Reasoning
To Students and Teachers

Residents of the United States face critical quantitative reasoning (QR) challenges more often than residents of any society in history. These challenges occur for two major reasons:

- Personal prosperity in US society requires numerous QR-based decisions.
- Sustaining US democratic processes requires citizens who can reason quantitatively.

These two demands for QR are required of all US citizens and are in addition to the QR demands of the workplace, which vary depending on the area of work.

This casebook provides a tool for educational response to the enormous QR demands that US residents face. It is the foundation for developing and delivering an ever-fresh, real-world-based course that starts or moves students forward on a path toward quantitative literacy. The terms quantitative literacy (QL) and quantitative reasoning (QR) are used interchangeably, and other terms such as numeracy essentially have the same meaning.

The contents of this book are governed by two criteria:

- Every QR problem is a contextual problem. This means that the quantitative reasoning is about a circumstance that is embedded in or grows out of a real-world context.
- Every mathematical or statistical topic investigated is one that is contained in or is useful in critiquing a public media article.

This book contains thirty case studies of public media articles and advertisements, mostly from newspapers. Also included are introductory notes and exercises on the basic concepts of understanding and comparing quantities; percent and percent change; indices; interest on money; weighted averages; reading and critiquing graphs; counting; and probability, odds, and risk. Each of the articles contains quantitative information, analyses, or argument. These case studies are meant to be both items of study and examples of case studies that students and teachers can create using public media articles from the present day, keeping the material fresh and more obviously relevant. Students discovering and presenting articles for class discussion that illustrate QR increase essential student engagement in the course. By creating case studies and discussing those in this book, students will develop

> Quantitative literacy (QL) is a habit of mind, and, consequently, achieving QL requires both extensive interaction between students and teachers and practice beyond school. At the collegiate level, we are concerned with a high level of QL, befitting persons with baccalaureate degrees, analogous to what Lawrence Cremin (1988) termed liberating literacy, as opposed to inert literacy. Therefore, the QL we seek includes command of both the enabling skills needed to search out quantitative information and power of mind necessary to critique it, reflect upon it, and apply it in making decisions.
>
> Cremin, L. A. (1988). *American education: The Metropolitan experience 1876–1980*. New York, NY: Harper & Row.

reasoning skills and disposition toward continuing to practice those skills beyond a course and beyond school. They will develop a habit of mind to reason quantitatively in their everyday lives as citizens, consumers, and workers.

A course based on this book will differ in many ways from traditional courses in mathematics or statistics. First, the book is not organized by mathematical topic, rather by the reasoning domains required to understand and critique articles in public media. Second, the exercises and study questions are not versions of template problems organized by solution methods. Third, and most important, the exercises and study questions are designed to address common, challenging quantitative reasoning situations in daily life.

This approach to teaching toward quantitative literacy (QL) is based on the assumption that there are canonical QL situations that students need to address and resolve. After encountering the situations, resolving them often involves the following steps:

- Interpreting the circumstances of the situation, making estimates as necessary to decide what investigation or study is merited and gleaning critical information.
- Representing the information in an appropriate mathematical form.
- Performing calculations and manipulations on the mathematical form.
- Analyzing or synthesizing the quantitative information produced.
- Evaluating assumptions that have been made.
- Communicating the results by reflecting back into the original circumstance.

These steps often require careful reading of both continuous prose and discontinuous prose (such as graphical representations), using mathematics or statistics, and then interpreting and critiquing the original prose in light of the mathematical results. Critical reasoning is required throughout. In general, students are not expecting this complicated process because their previous mathematics experiences have been narrower and more precisely defined. Frequently, the third phase, calculating, gets the most attention because it is the principal process of traditional mathematics and statistics courses.

Many of the problems are ill posed and require reasonable assumptions to resolve, and many of the problems have multiple reasonable responses. Consequently, conclusions require explanations of reasoning that led to the conclusions. Students should always read critically, understand a situation, and draw conclusions based on evidence. Communicating the results, including the evidence that underlies the results is the critical end product. So, reading and writing—communication—is a substantial portion of quantitative reasoning.

To Teachers

Quantitative Reasoning Proficiency

The model of mathematical proficiency as described in the National Research Council publication *Adding It Up*[1] is helpful in understanding the challenges of QR. In the model, mathematical proficiency has five intertwined strands (See the accompanying box).

Whereas some traditional mathematics courses depended strongly on the second strand, procedural fluency, all five of these strands are critical for QR, and the last two listed, adaptive reasoning and productive disposition, take on added importance for proficiency in QR. This stems from the fact that QR is a habit of mind and requires adapting reasoning to numerous unpredictable contexts. Productive disposition is often missing from students and non-students when it comes to understanding and critiquing quantitative material. The complaints by students, "where will I ever use this" and "I was never good at math," run counter to productive disposition because they question the relevance of school mathematics and one's own efficacy to understand and use it.

> Mathematical Proficiency from *Adding It Up*
>
> * conceptual understanding—comprehension of mathematical concepts, operations and relations
> * procedural fluency—skill in carrying out procedures flexibly, accurately, efficiently, and appropriately
> * strategic competence—ability to formulate, represent, and solve mathematical problems
> * adaptive reasoning—capacity for logical thought, reflection, explanation, and justification
> * productive disposition—habitual inclination to see mathematics as sensible, useful, and worthwhile, coupled with a belief in diligence and one's own efficacy.

The five strands of mathematical proficiency along with the process standards of the National Council for Teachers of Mathematics (NCTM) (problem solving, reasoning and proof, communication, connections, and representation) are the basis for the eight mathematical practice standards of the Common Core State Standards for Mathematics (CCSSM), which have been adopted by almost all states. Although these practice standards are aimed at K–12 mathematics, this CCSSM model of mathematical proficiency is very helpful in guiding student development throughout any mathematical reasoning course.

CCSSM Standards for Mathematical Practice

The Common Core State Standards for Mathematical Practice describe varieties of expertise that mathematics educators at all levels should seek to develop in their students. These practices rest on important "processes and proficiencies" with longstanding importance in mathematics education. The first of these are the NCTM process standards of problem solving, reasoning and proof, communication, representation, and connections. The second are the strands of mathematical proficiency specified in the National Research Council's report Adding It Up: adaptive reasoning, strategic competence, conceptual understanding, procedural fluency, and productive disposition.

---

1. Kilpatrick, J., Swafford, J., & Findell, B., Eds. (2001). *Adding it up*. Washington, DC: National Academies Press.

1. Make sense of problems and persevere in solving them.

Mathematically proficient students start by explaining to themselves the meaning of a problem and looking for entry points to its solution. They analyze givens, constraints, relationships, and goals. They make conjectures about the form and meaning of the solution and plan a solution pathway rather than simply jumping into a solution attempt. They consider analogous problems, and try special cases and simpler forms of the original problem in order to gain insight into its solution. They monitor and evaluate their progress and change course if necessary. Older students might, depending on the context of the problem, transform algebraic expressions or change the viewing window on their graphing calculator to get the information they need. Mathematically proficient students can explain correspondences between equations, verbal descriptions, tables, and graphs or draw diagrams of important features and relationships, graph data, and search for regularity or trends. Younger students might rely on using concrete objects or pictures to help conceptualize and solve a problem. Mathematically proficient students check their answers to problems using a different method, and they continually ask themselves, "Does this make sense?" They can understand the approaches of others to solving complex problems and identify correspondences between different approaches.

2. Reason abstractly and quantitatively.

Mathematically proficient students make sense of quantities and their relationships in problem situations. They bring two complementary abilities to bear on problems involving quantitative relationships: the ability to decontextualize—to abstract a given situation and represent it symbolically and manipulate the representing symbols as if they have a life of their own, without necessarily attending to their referents—and the ability to contextualize, to pause as needed during the manipulation process in order to probe into the referents for the symbols involved. Quantitative reasoning entails habits of creating a coherent representation of the problem at hand; considering the units involved; attending to the meaning of quantities, not just how to compute them; and knowing and flexibly using different properties of operations and objects.

3. Construct viable arguments and critique the reasoning of others.

Mathematically proficient students understand and use stated assumptions, definitions, and previously established results in constructing arguments. They make conjectures and build a logical progression of statements to explore the truth of their conjectures. They are able to analyze situations by breaking them into cases, and can recognize and use counterexamples. They justify their conclusions, communicate them to others, and respond to the arguments of others. They reason inductively about data, making plausible arguments that take into account the context from which the data arose. Mathematically proficient students are also able to compare the effectiveness of two plausible arguments, distinguish correct logic or reasoning from that which is flawed, and—if there is a flaw in an argument—explain what it is. Elementary students can construct arguments using concrete referents such as objects, drawings, diagrams, and actions. Such arguments can make sense and be correct, even though they are not generalized or made formal until later grades. Later, students learn to determine domains to which an argument applies. Students at all grades can listen or read the arguments of others, decide whether they make sense, and ask useful questions to clarify or improve the arguments.

---

4. Model with mathematics.

Mathematically proficient students can apply the mathematics they know to solve problems arising in everyday life, society, and the workplace. In early grades, this might be as simple as writing an addition equation to describe a situation. In middle grades, a student might apply proportional reasoning to plan a school event or analyze a problem in the community. By high school, a student might use geometry to solve a design problem or use a function to describe how one quantity of interest depends on another. Mathematically proficient students who can apply what they know are comfortable making assumptions and approximations to simplify a complicated situation, realizing that these may need revision later. They are able to identify important quantities in a practical situation and map their relationships using such tools as diagrams, two-way tables, graphs, flowcharts and formulas. They can analyze those relationships mathematically to draw conclusions. They routinely interpret their mathematical results in the context of the situation and reflect on whether the results make sense, possibly improving the model if it has not served its purpose.

---

5. Use appropriate tools strategically.

Mathematically proficient students consider the available tools when solving a mathematical problem. These tools might include pencil and paper, concrete models, a ruler, a protractor, a calculator, a spreadsheet, a computer algebra system, a statistical package, or dynamic geometry software. Proficient students are sufficiently familiar with tools appropriate for their grade or course to make sound decisions about when each of these tools might be helpful, recognizing both the insight to be gained and their limitations. For example, mathematically proficient high school students analyze graphs of functions and solutions generated using a graphing calculator. They detect possible errors by strategically using estimation and other mathematical knowledge. When making mathematical models, they know that technology can enable them to visualize the results of varying assumptions, explore consequences, and compare predictions with data. Mathematically proficient students at various grade levels are able to identify relevant external mathematical resources, such as digital content located on a website, and use them to pose or solve problems. They are able to use technological tools to explore and deepen their understanding of concepts.

---

6. Attend to precision.

Mathematically proficient students try to communicate precisely to others. They try to use clear definitions in discussion with others and in their own reasoning. They state the meaning of the symbols they choose, including using the equal sign consistently and appropriately. They are careful about specifying units of measure, and labeling axes to clarify the correspondence with quantities in a problem. They calculate accurately and efficiently, express numerical answers with a degree of precision appropriate for the problem context. In the elementary grades, students give carefully formulated explanations to each other. By the time they reach high school they have learned to examine claims and make explicit use of definitions.

---

7. Look for and make use of structure.

Mathematically proficient students look closely to discern a pattern or structure. Young students, for example, might notice that three and seven more is the same amount as seven and three more, or they may sort a collection of shapes according to how many sides the shapes have. Later, students will see 7×8 equals the well-remembered $7 \times 5 + 7 \times 3$, in preparation for learning about the distributive property. In the expression $x^2 + 9x + 14$, older students can see the 14 as 2×7 and the 9 as $2 + 7$. They recognize the significance of an existing line in a geometric figure and can use the strategy of drawing an auxiliary line for solving problems. They also can step back for an overview and shift perspective. They can see complicated things, such as some algebraic expressions, as single objects or as being composed of several objects. For example, they can see $5 - 3(x - y)^2$ as 5 minus a positive number times a square and use that to realize that its value cannot be more than 5 for any real numbers x and y.

8. Look for and express regularity in repeated reasoning.
Mathematically proficient students notice if calculations are repeated, and look both for general methods and for shortcuts. Upper elementary students might notice when dividing 25 by 11 that they are repeating the same calculations over and over again, and conclude they have a repeating decimal. By paying attention to the calculation of slope as they repeatedly check whether points are on the line through (1, 2) with slope 3, middle school students might abstract the equation $(y - 2)/(x - 1) = 3$. Noticing the regularity in the way terms cancel when expanding $(x - 1)(x + 1)$, $(x - 1)(x^2 + x + 1)$ and $(x - 1)(x^3 + x^2 + 1)$ might lead them to the general formula for the sum of a geometric series. As they work to solve a problem, mathematically proficient students maintain oversight of the process, while attending to the details. They continually evaluate the reasonableness of their intermediate result.

Content of Casebook

As mentioned above, there are six different types of materials in this book:

1) Introductory notes on basic concepts;
2) Warm up exercises on the basic concepts involved in case studies;
3) Articles that are the subject of case studies;
4) Study questions on the articles;
5) Suggestions for projects and papers;
6) Additional exercises for each section (in back of the book).

The questions and tasks take different forms, including the following:

a. Identifying and reporting quantitative information and arguments from the articles.
b. Developing additional quantitative information from information in the articles.
c. Critiquing the arguments, analyses, and conclusions of the articles.
d. Extending the arguments beyond those in the articles.
e. Research and reporting on concepts related to the articles.

The questions and tasks in the case studies of the articles can be used in various ways, such as for the basis of class discussion; as homework assigned for student responses; or to be discussed and completed by groups of students in class to produce a consensus group response.

Order of Material

The content is arranged in six sections, sorted by basic concepts that occur in the articles, but various concepts recur throughout the case studies. Occasionally, concepts that are unfamiliar to students are encountered without full explanation. For example, it is assumed that students can produce graphs of linear and exponential functions. Section 1 on Using Numbers and Quantities is fundamental to personal quantitative reasoning. Comparing and understanding quantities, especially large quantities, continue as themes throughout several of the case studies. The ideas of Section 2 on Percent and Percent Change also continue to recur throughout. Otherwise, there is little interdependence of the sections and cases, so one may change the order, skip case studies, or skip entire sections.

Prerequisites

Proportional, graphical, statistical, and algebraic reasoning are required to analyze the cases in this book and numerous similar cases that can be developed by students and teachers. Basic knowledge of algebra, descriptive statistics, and proportionality is necessary, but there is little dependence on algorithms and complex mathematical concepts. No knowledge of trigonometry, analytic geometry, or calculus is assumed, but the ideas of all three (proportionality, geometric reasoning, rate of change, approximation, etc.) are very helpful in fully developing the study of various cases. In terms of course prerequisites, students need to have a working knowledge of middle school mathematics and high school algebra (or college algebra). The sophistication of the case studies derives mostly from the contexts that span economics, sociology, politics, government policies, entertainment, health, and measurement.

Website

Some materials that support using these case studies are posted on the Quantitative Reasoning in the Contemporary World website at http://www.cwu.edu/~boersmas/QRCW/. Sample syllabi, discussion of quantitative reasoning, and links to related materials are available to anyone, and sample answers to the study questions are available but are password protected. The password is available to instructors from the authors. A mapping between the study questions and the core competencies of the Quantitative Literacy Assessment Rubric (discussed below) is also available.

Assessment

The nature of the exercises and study questions in this casebook requires new thought about assessing student work. That which is valued in a response to a QR problem situation differs from that which is valued in traditional disciplinary-based courses, especially those in mathematics or statistics. Assessing QR involves judging written analyses and reflections and the quality of evidence given in support of arguments or conclusions. Because of these differences, frequent reminders are necessary for students to supply the following:

- Evidence supporting reasoning or assertions.
- Calculations that produce numerical results.
- Correct units on quantities.
- Complete and correct sentences stating evidence and conclusions.
- Precision of language in stating questions and results.

One result of the authors' experiences with assessing student work was the development of a rubric[2] for this purpose. The rubric is given below and is helpful in guiding the development of student thinking, posing assessment questions, and scoring student work.

---

2. Boersma, Stuart; Caren Diefenderfer; Shannon W. Dingman; and Bernard L. Madison. 2011. Quantitative Reasoning in the Contemporary World, 3: Assessing Student Learning. *Numeracy*, 4 (2): Article 8. http://services.bepress.com/numeracy/vol4/iss2/art8

Quantitative Literacy Assessment Rubric

History

The Association of American College and Universities (AAC&U) released a Quantitative Literacy VALUE Rubric in 2009, which was designed by a team of college and university faculty from across the United States. The QL VALUE Rubric was one of 15 VALUE rubrics designed around a variety of learning outcomes. The intent of the designers was to provide an instrument for faculty to assess QL achievement in different types of student work, or collections of work. We have modified slightly the core competencies (names and definitions) and have rewritten the milestone descriptors. In order to avoid confusion with AAC&U's original rubric, we call the resulting instrument the "Quantitative Literacy Assessment Rubric."

Scope and Use

The quantitative literacy we seek to assess includes command of both the enabling skills needed to search out quantitative information and the power of mind necessary to critique it, reflect upon it, and apply it in making decisions.

The QL Assessment Rubric is intended to measure achievement levels of the associated QL core competencies in a variety of student work. Depending on the nature of the prompt, not all competencies will be present in every student response. Occasionally there may be a perceived overlap of the QL core competencies. In which case, one competency should be chosen as the dominant one to assess. Careful wording of the prompt is often crucial to obtaining detailed student responses. A mapping between the study questions and the core competencies of the Quantitative Literacy Assessment Rubric may be found on the project website: http://www.cwu.edu/~boersmas/QRCW.

Quantitative Literacy Core Competency	Achievement Level			
	3	2	1	0
Interpretation *Ability to glean and explain mathematical information presented in various forms (e.g. equations, graphs, diagrams, tables, words)*	Correctly identifies all relevant information.	Correctly identifies some, but not all, relevant information.	Some relevant information is identified, but none is correct.	No relevant information identified.
Representation *Ability to convert information from one mathematical form (e.g. equations, graphs, diagrams, tables, words) into another.*	All relevant conversions are present and correct.	Some correct and relevant conversions are present but some conversions are incorrect or not present.	Some information is converted, but it is irrelevant or incorrect.	No conversion is attempted.
Calculation *Ability to perform arithmetical and mathematical calculations.*	Calculations related to the problem are correct and lead to a successful completion of the problem.	Calculations related to the problem are attempted but either contain errors or are not complete enough to solve the problem.	Calculations related to the problem are attempted but contain errors and are not complete enough to solve the problem.	Calculations given are not related to the problem, or no work is present.
Analysis/Synthesis *Ability to make and draw conclusions based on quantitative analysis.*	Uses correct and complete quantitative analysis to make relevant and correct conclusions.	Quantitative analysis is given to support a relevant conclusion but it is either only partially correct or partially complete (e.g. there are logical errors or unsubstantiated claims).	An incorrect quantitative analysis is given to support a conclusion.	Either no reasonable conclusion is made or, if present, is not based on quantitative analysis.
Assumptions *Ability to make and evaluate important assumptions in estimation, modeling, and data analysis.*	All assumptions needed are present and justified when necessary.	At least one correct and relevant assumption is given (perhaps coupled with erroneous assumptions), yet some important assumptions are not present.	Attempts to describe assumptions, but none of the assumptions described are relevant.	No assumptions present.
Communication *Ability to explain thoughts and processes in terms of what evidence is used, how it is organized, presented, and contextualized.*	A correct and complete explanation is clearly presented.	A partially correct relevant explanation is present, but incomplete or poorly presented.	A relevant explanation is present, but is illogical, incorrect, illegible, or incoherent.	No relevant explanation is provided.

Acknowledgements

We gratefully acknowledge support of the National Science Foundation for the further development and expansion of the quantitative reasoning course and the production of this book (DUE-0715039).

Thirteen advisors for the NSF project, Quantitative Reasoning in the Contemporary World (QRCW), have provided continuing valuable counsel as this book has evolved. Some advisors were from the University of Arkansas: Robert Babcock (Engineering), Daniel Ferritor (Sociology), Andrew Gibbs (Drama), John Hehr (Geography), Gay Stewart (Physics), and Patsy Watkins (Journalism). Advisors from other institutions were James Choike (Mathematics, Oklahoma State University), Joan Leitzel (Mathematics, University of New Hampshire), Barbara Moskal (Mathematics Education, Colorado School of Mines), Milo Schield (Statistics, Augsburg College), Lynn Steen (Mathematics, St. Olaf College), Corrine Taylor (Economics, Wellesley College), and Alan Tucker (Applied Mathematics, State University of New York at Stony Brook). The evaluator for the NSF project was Terri J. Murphy (Northern Kentucky University).

We are also pleased to acknowledge the contributions of instructors of the QRCW course at the University of Arkansas, especially Amber Bocquin and Tyler Fuller, and the middle and high school science and mathematics teachers in summer workshops in quantitative literacy who used versions of this book to explore ways to connect their classes more closely to the quantitative reasoning demands of contemporary US society.

Bernard L. Madison
University of Arkansas
Fayetteville, Arkansas

Shannon W. Dingman
University of Arkansas
Fayetteville, Arkansas

Stuart Boersma
Central Washington University
Ellensburg, Washington

Caren L. Diefenderfer
Hollins University
Roanoke, Virginia

Quantitative Reasoning

Section 1

Using Numbers and Quantities

In Section 1 we encounter several uses and misuses of numbers, particularly in news articles. The goal is to introduce some concepts (for example, comparing quantities, percent and percent change) that we will study more in depth in later lessons. The content of this section is below.

- Introduction to numbers and quantities.
- Case Study 1.1: "What $1.2 Trillion Can Buy" by David Leonhardt, *New York Times,* January 17, 2007.
- Case Study 1.2: "Numbed by the Numbers, When They Just Don't Add Up" by Daniel Okrent, *New York Times,* January 23, 2005, and Web Posting No. 42, "Three Bad Numbers" by Daniel Okrent, *New York Times.*
- Case Study 1.3: "UA passes produce a remarkable ratio" by Bob Holt, *Arkansas Democrat-Gazette*, November 22, 2006.
- Case Study 1.4: Harps Food Stores Advertising Poster.

Introduction to Numbers and Quantities

The learning goals of this section include the following:

- Understanding how numbers or quantities are used and misused
- Computing on the fly: mental calculations and approximations
- Comparing numbers or quantities
- Understanding large quantities in personally meaningful terms (units)
- Understanding the importance of correct units

These learning goals will continue to be central throughout the cases in this book, but they are explored here with four case studies.

Number and quantity

Number and quantity are different. A quantity is a number with a unit, and units are very important in quantitative reasoning. It is helpful to think of numbers as adjectives and units as nouns. For example, 2 pounds of apples is a quantity, but 2 is a number. Most solutions to quantitative reasoning problems are quantities, and correct and complete units are essential parts of the solutions.

Computing on the fly

Often when reading a media article that contains quantitative information and quantitative analyses or arguments, one does not have the time or interest to check quantitative assertions with precision. Consequently, it is expedient to check assertions with mental calculations and approximations. For example, if an article claims that a quantity has increased by 300 percent then mentally one can check if the new value of the quantity is 4 times its earlier value. Or if an article claims that 1.5 percent of the population of the US resides in Pittsburgh, PA, one needs to be able to compute 1.5% of the US population. Note that 1% of 300 million is 3 million and another half percent is 1.5 million. So 1.5% of the US population is about 4.5 million, which is far more than the population of Pittsburgh, which was approximately 305,000 in 2010.

Comparing numbers or quantities

There are several standard ways to compare quantities and numbers.

1) <u>More/less than:</u> Six is more than 4. Three pounds of apples is less than 4 pounds of apples. Is 3 pounds of apples less than 4 pounds of oranges?
2) <u>How much more/less than:</u> Six is 2 more than 4. Six bushels of potatoes is 2 bushels more than 4 bushels of potatoes.
3) <u>Ratio:</u> Ten is twice 5. Five is half of ten.
4) <u>Percent (modified ratio):</u> Ten is 200% of 5. Five is 50% of 10.
5) <u>Percent more/percent less:</u> Ten is 25% more than 8. Eight is 20% less than 10.
6) <u>Times more/times less:</u> What does it mean to say that a number is "ten times more than 5"? What does it mean to say that a number is "ten times less than 5"? Although both "times more" and "times less" are used in public discourse, their meanings are confusing. The confusion partly stems from the object of "times."

Ten times 5 is 50, so ten times more than 5 could be interpreted as 55. Often the intended meaning of ten times more than 5 is the same as ten times 5 or 50. The expression "ten times less" is more difficult to interpret; one reasonable meaning is "one-tenth" times.

Notice that comparing quantities in different units is usually meaningless. For example, what does it mean to say that "5 oranges" is more than "3 apples?" Often, as is the case here, one can compare quantities by finding a common unit. Here, one common unit is pieces of fruit.

Understanding large quantities

How much is $1.2 trillion or $878 billion? How far is it to the moon or the sun? Billions and trillions of dollars are not easily understood in units that are personally meaningful. Will $1.2 trillion purchase a $150,000 home for every person in the state of New York? How long would it take to travel to the moon at the speed of 600 miles per hour, the speed of a passenger jet airplane? To understand these large quantities, one often expresses them in units, or terms, that are more meaningful. For example, if the house you are currently living in cost $150,000, you may wish to express large dollar amounts such as $878 billion in terms of $150,000 houses. There is a dual problem of understanding small numbers or measurements, but in this lesson we look at large quantities because they occur more often in media articles.

Understanding the importance of correct units

A quantity is a number with an accompanying unit. Common examples of quantities include 6 apples, 5 dollars, 20 euros, 12.5 pounds, 1.72 meters, and 2.68 inches. Comparing or adding quantities requires that they have the same units. For example, comparing $4 and 3 euros requires knowing a conversion factor and converting both quantities into dollars or into euros (or both into some other unit such as pesos). Once both quantities have the same units, we then can compare them or compute with the given values. Another example involves units of measuring lengths: 18 inches is more than 1.2 feet because 18 inches is 1.5 feet and 1.5 feet is larger than 1.2 feet. Some quantities simply do not compare. You have probably heard the saying, "comparing apples to oranges." Likewise, 3 apples + 4 oranges is neither 7 apples nor 7 oranges. However, there is a common unit: pieces of fruit. There is an analogy to this in adding fractions: $\frac{2}{3} + \frac{5}{7}$ can be viewed as adding two one-thirds and five one-sevenths. By finding a common denominator for both fractions, we find a common unit in order to compute with or compare the fractions. A common unit for these fractions is twenty-firsts: $\frac{2}{3} = \frac{14}{21}$ and $\frac{5}{7} = \frac{15}{21}$ and so $\frac{2}{3} + \frac{5}{7} = \frac{14}{21} + \frac{15}{21} = \frac{29}{21}$. Also, one can now compare $\frac{2}{3}$ and $\frac{5}{7}$. Since $\frac{14}{21} < \frac{15}{21}$, we know that $\frac{2}{3} < \frac{5}{7}$.

As we will see throughout the cases in this book, measuring quantities in different units is often the basis for quantitative arguments. Furthermore, very few of the answers to questions posed in this book are numbers. They are numbers with units. Consequently, for each numerical answer such as 7.5 one should ask 7.5 what?

Case Study 1.1: What $1.2 Trillion Can Buy

Resource Materials: "What $1.2 Trillion Can Buy" by David Leonhardt, *New York Times*, January 17, 2007.

Learning Goals: The learning goals of this case study include understanding large quantities by expressing them in more familiar terms and comparing quantities in various ways.

Warm Up Exercises for Case Study 1.1

For numbers 1–4, find each distance in miles, and give the source of your information.

1. Find the distance from New York, NY, to San Francisco, CA, via I-80.
2. Find the distance from Earth to the Moon.
3. Find the distance from Earth to the Sun.
4. Find the diameter of the Earth at the Equator.
5. Write a sentence to compare the distances in #2 and #3 using a ratio, as in "The distance from the Earth to the Sun is _____ times the distance from the Earth to the Moon."
6. Write a sentence to compare the distances in #2 and #3 using percent, as in "The distance from the Earth to the Moon is _____ % of the distance from the Earth to the Sun."
7. Write a sentence to compare the distances in #1 and #4 using a ratio.
8. Write a sentence to compare the distances in #1 and #4 using a percent.
9. Provide an explanation to a friend that would help him/her understand the distance from the Earth to the Moon in terms he/she understands.

Article for Case Study 1.1

The New York Times
January 17, 2007
Economix
What $1.2 Trillion Can Buy
By DAVID LEONHARDT

The human mind isn't very well equipped to make sense of a figure like $1.2 trillion. We don't deal with a trillion of anything in our daily lives, and so when we come across such a big number, it is hard to distinguish it from any other big number. Millions, billions, a trillion—they all start to sound the same.

The way to come to grips with $1.2 trillion is to forget about the number itself and think instead about what you could buy with the money. When you do that, a trillion stops sounding anything like millions or billions.

For starters, $1.2 trillion would pay for an unprecedented public health campaign—a doubling of cancer research funding, treatment for every American whose diabetes or heart disease is now going unmanaged and a global immunization campaign to save millions of children's lives.

Combined, the cost of running those programs for a decade wouldn't use up even half our money pot. So we could then turn to poverty and education, starting with universal preschool for every 3- and 4-year-old child across the country. The city of New Orleans could also receive a huge increase in reconstruction funds.

The final big chunk of the money could go to national security. The recommendations of the 9/11 Commission that have not been put in place—better baggage and cargo screening, stronger measures against nuclear proliferation—could be enacted. Financing for the war in Afghanistan could be increased to beat back the Taliban's recent gains, and a peacekeeping force could put a stop to the genocide in Darfur.

All that would be one way to spend $1.2 trillion. Here would be another:

The war in Iraq.

In the days before the war almost five years ago, the Pentagon estimated that it would cost about $50 billion. Democratic staff members in Congress largely agreed. Lawrence Lindsey, a White House economic adviser, was a bit more realistic, predicting that the cost could go as high as $200 billion, but President Bush fired him in part for saying so.

These estimates probably would have turned out to be too optimistic even if the war had gone well. Throughout history, people have typically underestimated the cost of war, as William Nordhaus, a Yale economist, has pointed out.

But the deteriorating situation in Iraq has caused the initial predictions to be off the mark by a scale that is difficult to fathom. The operation itself—the helicopters, the tanks, the fuel needed to run them, the combat pay for enlisted troops, the salaries of reservists and contractors, the rebuilding of Iraq—is costing more than $300 million a day, estimates Scott Wallsten, an economist in Washington.

That translates into a couple of billion dollars a week and, over the full course of the war, an eventual total of $700 billion in direct spending.

The two best-known analyses of the war's costs agree on this figure, but they diverge from there. Linda Bilmes, at the Kennedy School of Government at Harvard, and Joseph Stiglitz, a Nobel laureate and former Clinton administration adviser, put a total price tag of more than $2 trillion on the war. They include a number of indirect costs, like the economic stimulus that the war funds would have provided if they had been spent in this country.

Mr. Wallsten, who worked with Katrina Kosec, another economist, argues for a figure closer to $1 trillion in today's dollars. My own estimate falls on the conservative side, largely because it focuses on the actual money that Americans would have been able to spend in the absence of a war. I didn't even attempt to put a monetary value on the more than 3,000 American deaths in the war.

Besides the direct military spending, I am including the gas tax that the war has effectively imposed on American families (to the benefit of oil-producing countries like Iran, Russia and Saudi Arabia). At the start of 2003, a barrel of oil was selling for $30. Since then, the average price has been about $50. Attributing even $5 of this difference to the conflict adds another $150 billion to the war's price tag, Ms. Bilmes and Mr. Stiglitz say.

The war has also guaranteed some big future expenses. Replacing the hardware used in Iraq and otherwise getting the United States military back into its prewar fighting shape could cost $100 billion. And if this war's veterans receive disability payments and medical care at the same rate as veterans of the first gulf war, their health costs will add up to $250 billion. If the disability rate matches Vietnam's, the number climbs higher. Either way, Ms. Bilmes says, "It's like a miniature Medicare."

In economic terms, you can think of these medical costs as the difference between how productive the soldiers would have been as, say, computer programmers or firefighters and how productive they will be as wounded veterans. In human terms, you can think of soldiers like Jason Poole, a young corporal profiled in The New York Times last year. Before the war, he had planned to be a teacher. After being hit by a roadside bomb in 2004, he spent hundreds of hours learning to walk and talk again, and he now splits his time between a community college and a hospital in Northern California.

Whatever number you use for the war's total cost, it will tower over costs that normally seem prohibitive. Right now, including everything, the war is costing about $200 billion a year.

Treating heart disease and diabetes, by contrast, would probably cost about $50 billion a year. The remaining 9/11 Commission recommendations—held up in Congress partly because of their cost—might cost somewhat less. Universal preschool would be $35 billion. In Afghanistan, $10 billion could make a real difference. At the National Cancer Institute, annual budget is about $6 billion.

This war has skewed our thinking about resources, said Mr. Wallsten, a senior fellow at the Progress and Freedom Foundation, a conservative-leaning research group. "In the context of the war, $20 billion is nothing."

As it happens, $20 billion is not a bad ballpark estimate for the added cost of Mr. Bush's planned surge in troops. By itself, of course, that price tag doesn't mean the surge is a bad idea. If it offers the best chance to stabilize Iraq, then it may well be the right option.

But the standard shouldn't simply be whether a surge is better than the most popular alternative—a far-less-expensive political strategy that includes getting tough with the Iraqi government. The standard should be whether the surge would be better than the political strategy *plus* whatever else might be accomplished with the $20 billion.

This time, it would be nice to have that discussion before the troops reach Iraq.

leonhardt@nytimes.com

Study Questions for Case Study 1.1

"What $1.2 Trillion Can Buy" by David Leonhardt
New York Times
January 17, 2007

This article discusses how to make sense of a large quantity, specifically $1.2 trillion. Understanding such a quantity depends heavily on what one understands beforehand, namely the cost of other items.

1. Of the writer's list of programs that could be funded with the $1.2 trillion, which two are the best for helping you understand this large amount of money? Why are the two you selected most helpful?

2. How many $150,000 homes would the $1.2 trillion buy? How can we make sense of this number of $150,000 homes? Determine an appropriate measure (e.g., population of a particular state or city) that would help someone make sense of the number of $150,000 homes that $1.2 trillion could buy.

3. Compare the writer's estimated cost of the war in Iraq to the Pentagon's estimate of the cost of the war using a ratio and a percent. Use each comparison in a sentence.

4. Compare the writer's estimated cost of the war to Lawrence Lindsey's estimate using a ratio and a percent. Use each comparison in a sentence.

5. Using Scott Wallsten's estimate, how many days of funding the war in Iraq would produce a dollar amount equal to the annual budget of the National Cancer Institute?

6. Choose some measure of the size of $1.2 trillion (other than homes) that helps you understand its magnitude and express the $1.2 trillion in your measure. The answer to #2 is an example of this. Explain why this measure is meaningful to you.

Case Study 1.2: Numbed by the Numbers & Three Bad Numbers

> Resource Materials: "Numbed by the Numbers, When They Just Don't Add Up" by Daniel Okrent, *New York Times*, January 23, 2005, and "Three Bad Numbers" by Daniel Okrent, *New York Times* Web posting.

Learning Goals: The learning goals of this case study include understanding how numbers and quantities are used and misused.

Warm Up Exercises for Case Study 1.2

1. Find 2% of 3,500,000.

2. Find a number that is 80% more than 600.

3. Find a number that is 20% less than 500.

4. Find the average (also called the mean) of 20, 45, 72, and 98.

5. If 2% of the people in the US have red hair and there are 610,000 people in the US with red hair, find the population of the US. Is your answer reasonable? Why or why not?

6. In baseball, a player's batting average is the three-place decimal that is the fraction of hits in the number of times at bat, where certain at-bats such as walks are not counted. For example, a player who has 5 hits in 20 at-bats has a batting average of .250. Find the batting average of a player who has 27 hits in 75 at-bats.

Articles for Case Study 1.2

New York Times
January 23, 2005
THE PUBLIC EDITOR
Numbed by the Numbers, When They Just Don't Add Up
By DANIEL OKRENT

Some people in the newspaper business—including, I suspect, a few sitting upstairs from me, in the New York Times Company's corporate offices—were displeased by a story that ran on Jan. 10, "Your Daily Paper, Courtesy of a Sponsor." The article, by Jacques Steinberg and Tom Torok, was a pretty sharp pin stuck into the circulation numbers of many American newspapers, revealing how subscriptions paid for by advertisers are delivered to readers who haven't asked for them.

I fielded a couple of days' worth of objections from the newspaper industry, and while I concluded that the piece was largely fair and entirely accurate (if somewhat overstated), I do think it could have been more candid about The Times's own practices. Readers who wanted to know how The Times fitted into this story didn't find out until (more likely, "unless") they made it to the 30th paragraph; the practices at The *Boston Globe*, owned by The New York Times Company, were unveiled in Paragraph 27. Even then the article was slightly less than forthcoming. By studying circulation patterns of Sunday papers, the article made The Times appear less reliant on these advertiser-subsidized subscriptions than it would have if the comparisons had been based on weekday circulation.

In fact, one could say there's a stark difference: according to the most recent available numbers, the quantity of the paper's third-party-paid subscriptions on a given weekday is 79 percent higher than the comparable Sunday number.

This sounds very ominous. It sounds somewhat less ominous when you realize that these same third-party-paid subscriptions account for 1.4 percent of Sunday circulation, and 2.5 percent of weekday circulation. And it sounds not even worth noting (take a deep breath here) if you consider that the difference between the number of weekday subsidized copies and Sunday subsidized copies is 0.4 percent of weekday circulation, and 0.27 percent of Sunday circulation.

Set aside the question of whether The Times should have stated its figures higher and more completely in the piece. (No, let's not set it aside: Caesar's wife should speak early and loudly.) There's another issue rolling around all these numbers—namely, numbers. Do you have any idea which of the figures I've cited, all of them accurate, are meaningful?

Neither do I.

One of the appealing things about the complaints I receive about innumeracy at The Times is their ecumenical origin; when it comes to how it handles numbers, The Times is an equal opportunity offender. Like a bad cough that spreads its germs indiscriminately, numbers misapplied and ill-explained irritate the sensibilities of the right and the left, the drug company official and the animal rights activist, the art collector and the Jets fan.

Number fumbling arises, I believe, not from mendacity but from laziness, carelessness or lack of comprehension. I'll put myself in the latter category (as some readers no doubt will as well, after they've read through my representation of the numbers that follow). Most of the journalists I know who enter the profession comfortable with numbers write about sports, where debate about the meaning of statistics is a daily competition, or economics, a field in which interpretation of numbers will no more likely produce inarguable results than will finger painting.

So it is left to the rest of us who write for the paper to stumble through numbers, scatter them on the page and hope that readers understand. Does it matter if many of these figures are meaningless symbols serving the interests of the parties that issue them? Take a variety of reports on some recent lawsuits: A man is suing the city for $20 million arising from charges, eventually dismissed, brought against him for kidnapping and sexual abuse. The mother of the football player Derrick Thomas, who died in 2000, is suing General Motors for $75 million. Villagers on an Indonesian island are suing Newmont Mining Corporation for $543 million. Not one of these numbers is grounded in anything more substantial than the imagination of a plaintiff's lawyer, but each is given the authority of print.

No different, really, was Wednesday's assertion that Bernard J. Ebbers, if convicted of all charges in the MCI-WorldCom accounting scandal, "could be sentenced to as much as 85 years," a formulation that bears no relationship to any conceivable outcome yet serves the prosecutor's public case very nicely.

Numbers issued by those measuring criminal enterprise ("In Mexico, drug trafficking is a $250-billion-a-year industry") or the economic impact of a new stadium ("Bloomberg said that he expected the arena to generate about $400 million a year through various economic activities") don't deserve to be published without challenge; it doesn't serve agencies who want to fight drug trafficking to underestimate the problem, nor can any politician support a development project without hyping its potential benefit.

Still, The Times persists. In November, when New York City Comptroller William Thompson released a study purporting to show that New Yorkers purchase more than $23 billion in counterfeit goods each year, The Times repeated the analysis as if it were credible. Quick arithmetic would have demonstrated that $23 billion would work out to roughly $8,000 per city household, a number ludicrous on its face. (In the Web version of this column, I've linked to an excellent dissection of Thompson's report, by freelance journalist Felix Salmon.)

Last Sunday, an article on the city's proposed $1.1 billion investment in three stadium projects cited the assertion by the president of the city's Economic Development Corporation that "for every dollar invested by the city in the three projects, taxpayers would get a return of $3.50 to $4.50 over 30 years." It didn't say that the same $1.1 billion invested in a 30-year Treasury bond would return $4 for every dollar invested, and a lot more reliably, too. (Credit where it's due: reporter Charles V. Bagli did note that the $1.1 billion could pay for 25 schools housing 600 students each.)

Sometimes the absence of a number is as deflating to an article's credibility as the presence of a deceptive one. Few articles noting that President Bush received more votes than any candidate in history also mentioned that more people voted against him than any candidate in history. Quoting Michael Moore's assertion that standing ovations in Greensboro, N.C., proved that "Fahrenheit 9/11" is "a red state movie" disregards the fact that metropolitan Greensboro has over 1.2 million people. You could probably find in a population that large enough people to give a standing O for a reading of the bylaws of the American Dental Association.

Of course both Moore and the reporter who wrote that piece operate in the movie business, where records are about as meaningful as promises. "Shrek 2" is not, as an article in The Times Magazine had it in November, "the third-highest-grossing movie of all time" if you consider inflation, it's not even in the Top 10 (and "Titanic" is far from No. 1). This record-mania has spread everywhere. "Record-high gas prices" summoned up last year weren't even close; at its summer peak, gas cost 80 cents a gallon less than it did in 1981. Says economics reporter David Leonhardt, "Treating 2004 dollars the same as 1981 dollars isn't much different from treating dollars the same as rupees. The fact that 10 is a bigger number than 9 doesn't make 10 rupees worth more than $9; nor does it make $10 from 2004 worth more than $9 from 1981." Inflation isn't the only culprit stalking the record books: "Record deficits" may not be records when they're expressed as a percentage of gross domestic product, a far more reasonable measure than any raw number.

Numbers without context, especially large ones with many zeros trailing behind, are about as intelligible as vowels without consonants. When Congress allocated $28.4 billion to the National Institutes of Health, was that a lot or a little? I'd certainly begin to have a sense of it if I knew that this came to 3 percent of all discretionary spending. When John Kerry proposed tax cuts of $420 billion over 10 years, was that a meaningful number? Tell me that it amounts to about $150 per person per year, and I can grasp it. When Harvard announced that it was allocating $2 million more to financial aid for poor students, bringing the total to $82 million a year, was it really being generous? Well, in 2004, $82 million was about six days' income from the Harvard endowment, and the heralded $2 million increase that prompted this fairly prominent article was the equivalent of what the endowment earned every 3 hours and 36 minutes.

If all these numbers make your eyes roll, then you're finding yourself in the same position as a lot of readers, and apparently a lot of reporters and editors as well. (I haven't even gotten into deceptive stats that have the patina of authority, like those three all-time champs, the Dow Jones Industrial Average, the unemployment rate and batting averages; if you're interested, I take a few swings at them in my Web journal, in Posting No. 42.)

Although everyone who writes for The Times is presumably comfortable with words, every sentence nonetheless goes through the hands of copy editors, highly trained specialists who can bring life to a dead paragraph or clarity to a tortured clause with a tap-tap here and a delete-insert there. But numbers, so alien to so many, don't get nearly this respect. The paper requires no specific training to enhance numeracy, and no specialists whose sole job is to foster it. David Leonhardt and Charles Blow, the deputy design director for news, have just begun to conduct occasional seminars on "Using and Misusing Numbers," and that's a start. But as I read the paper and try to dodge the context-absent numbers that are thrown about like shot-puts, I long for more.

In "Floater," his 1980 novel about life at a newsweekly, Calvin Trillin introduced the Rhymes-With man—a mysterious character locked in a padded room who is allowed out only to provide readers with parenthetical clues to the pronunciation of foreign words, like ratatouille ("rhymes with lotta hooey"). Maybe The Times could sign up several Number-Means people to help the staff—and the readers—through the sticky digits.

The public editor serves as the readers' representative. His opinions and conclusions are his own. His column appears at least twice monthly in this section.

Three Bad Numbers
By Daniel Okrent

Daniel Okrent was the public editor of The New York Times from December 2003 until May 2005. This is taken from http//topics.nytimes.com/top/opinion/readersopinions/forums/thepubliceditor/danielokrent/

Some numbers are so well established in the newspaper writer's automatic vocabulary that they've earned an aura of authority they don't deserve. The Dow Jones Industrial Average is no more meaningful than a hiccup. If I were Dow Jones & Company, I'd be thrilled to have my brand name repeated daily in hundreds of newspaper columns, broadcast reports and web mentions, but I'd also keep my fingers crossed. Knowing that someday the press will no longer be so easily suckered.

There are three key problems with the Dow. Its tiny selection of 30 stocks (a number established three quarters of a century ago, in an infinitely less complex market era) can't begin to represent the variety of investment instruments sold on American exchanges; it specifically excludes transportation and utility companies, which Dow tracks separately; and it's mathematically preposterous.

Daniel Gross explained why in an article published two years ago on Slate.com: "Every time one of the stocks in it moves up one dollar, the Dow moves up a set number of points. In the real world, a dollar move on a $100 stock has an entirely different meaning than a $1 move on a $10 stock—but the Dow regards them as equal."

That's sort of like saying a $1.50 tip on a ten dollar breakfast is the same as a $1.50 tip on a hundred dollar dinner. The Standard & Poor's 500 not only incorporates a much larger number and much broader range of securities, it weights its elements logically. Times editors would be wise to execute a permanent "replace all" in their skulls and in their stories, forever banishing the Dow to the same attic where Wall Street keeps the ticker tape.

The unemployment rate issued by the Department of Labor each month is another substandard standard. It's a meaningful measure of several things, but not what you and I mean by "unemployment." By excluding those who have stopped looking for work, it can tend to overstate the apparent health of the economy.

Dave Leonhardt, who teaches The *Times*'s recently initiated staff seminar on numbers, notes that "over the last three years, the unemployment rate has fallen while the percentage of adults with jobs has dropped. Neither of these two measures tells the story on its own." On its own, unemployment rate is an inexact piece of shorthand too often conveying something very different from what the writer intends, or what the reader perceives.

Then there's batting average, to which no one serious pays serious attention (except for agents representing ballplayers with high BAs).

For one thing, it suggests great differences between hitters when they barely exist: the ballplayer with a .300 average (the age-old measure of quality) gets one more hit every two weeks than a .275 hitter. More telling, there's not a manager in baseball (at least not one who can keep his job) who, all other measurements being equal, would rather have a .300 hitter who never walks than a .275 hitter who draws a walk every day. The point, after all, is getting on base, which On-Base Percentage (hits plus walks divided by plate appearances) measures, and BA does not.

All three of these imprecise and generally unhelpful numbers have, through over-use and under-explanation, become part of the language, but none means what it purports to mean. Even writers who know better will at times resort to them when they're too hurried, too lazy or too weary to search for alternatives or to pause for elaboration.

A few weeks ago, David Leonhardt—who moonlights from his usual gig as an economics reporter to write a biweekly column on sports statistics—sought to establish that outfielder Carlos Beltran hadn't had such a great season last year, and clinched his case by noting that Beltran's batting average of .267 was "good for 118th place in baseball."

No journalist I know understands numbers as well as David does, no one has taught me more about them, and few are as good-natured as he is. That's why I don't think he'll mind if I point out that in July, just six months before he used it to evaluate Beltran, Leonhardt described batting average as a "flawed measure of performance."

Such is the persistent power of a bad number—it can bring the best of us to our knees.

Study Questions for Case Study 1.2

"Numbed by the Numbers, When They Just Don't Add Up" by Daniel Okrent
New York Times
January 23, 2005

"Three Bad Numbers" by Daniel Okrent
New York Times **Web journal**

1. Complete the following.

   a. Explain how the author probably got the "79 percent higher" statement in the third paragraph of the *Numbed by the Numbers* article. Critique the method that you come up with.

   b. Discuss whether or not the "79 percent higher" statement in the paper is accurate. If not accurate, is there a way to state it precisely?

2. If the weekday circulation of the *New York Times* is 2 million, find the Sunday circulation.

3. The author describes the following five misuses of numbers:

   a. Meaningless numbers stated to serve some particular purpose.

   b. Use or non-use of numbers that lack credibility.

   c. Flawed comparison of numbers.

   d. Numbers without context.

   e. Numbers/statistics that have undeserved authority.

   Choose three of these misuses and for each one, find an example of the misuse given by the author and an example of your own that you have observed in everyday life.

4. The author indicates that federal deficit figures should be expressed as a percentage of gross domestic product. Why does this make sense?

5. Give the three key problems the author lists for the Dow Jones Industrial Average (DJIA).

6. As mentioned in the article, the unemployment rate is often misunderstood. The unemployment rate is defined as the number of unemployed persons divided by the number of people in the labor force (consisting of all employed and unemployed persons). The key here is that the number of unemployed people only includes those adults who are unemployed and seeking employment. Using these definitions answer the following questions:

   a. What could cause the unemployment rate to fall?

   b. What could cause the percentage of adults with jobs to drop?

   c. Could both happen at the same time? Explain.

Case Study 1.3 Football Passing Ratios

Resource Materials: "UA passes produce a remarkable ratio" by Bob Holt, *Arkansas Democrat-Gazette*, November 22, 2006.

Learning Goals: The learning goals of this case study include using ratios for comparisons and understanding how those ratios change. Students will learn to use different strategies to solve problems.

Warm Up Exercises for Case Study 1.3

1. The batting average of a baseball player is the ratio of the number of hits to the number of official times at bat, which does not include some outcomes such as walks and being hit by a pitch. Batting averages are usually expressed as a decimal to three places such as .355.

   a. Find a player's batting average if the player has 23 hits in 97 official times at bat.

   b. The player from part (a) continues beyond the 97 official at bats with 23 hits and gets 5 hits in the next 12 at bats. Find the new batting average.

   c. The player from part (b) goes on a streak where he gets one hit for every three official at bats. Assuming this streak continues, find the minimum number of hits he will have that results in a batting average of .280.

2. Gregory, driving a truck and pulling a boat on a trailer, leaves Pines' Meadow campground for the long trip home. With the boat and trailer Gregory travels at 40 miles per hour.

   a. Gregory's sister, Alyson, leaves the campground an hour later and travels 50 miles per hour, following the same route as Gregory. If these two cars continue in this way, how long will it take Alyson to catch up with Gregory?

   b. Assuming that Gregory must drive at 40 miles per hour and Alyson starts an hour later as above, how fast must she travel in order to catch up with Gregory at some point down the road? For example, will Alyson catch up with Gregory if she travels 42 miles per hour?

Article for Case Study 1.3

UA passes produce a remarkable ratio
BY BOB HOLT
ARKANSAS DEMOCRAT-GAZETTE

FAYETTEVILLE—Arkansas is hitting home runs in the passing game at a Babe Ruthian pace this season.

Ruth hit 714 home runs in 8,398 career at-bats for an 11.8 ratio.

Arkansas also has an 11.8 ratio of touchdown passes to pass attempts through 11 games.

While Arkansas has attempted fewer passes (224) than any other SEC team and ranks 11th out of 12 conference teams in passing offense (156 yards per game), the Razorbacks have been productive when they have thrown.

Arkansas' 19 touchdown passes rank sixth in the SEC and the Razorbacks' ratio of touchdowns per pass attempts is the conference's best.

"We can air it when we need to do that," said junior wide receiver Marcus Monk, who leads Arkansas with nine touchdown catches.

LSU, the Razorbacks' opponent Friday in Little Rock, has the SEC's second-best ratio at 12.0 with 26 touchdowns in 312 pass attempts.

MAKING THEM COUNT

Arkansas has attempted fewer passes (224) than any other team in the SEC and ranks 11th in passing offense (156 yards per game), but the Razorbacks have been productive when they have thrown the ball. Arkansas ranks sixth in the SEC in touchdown passes (19) and is averaging a touchdown on every 11.8 attempts for the best ratio in the conference. Here is a rundown of how each SEC team ranks in touchdown passes per attempt:

TEAM	ATT.	TD PASSES	TDS/ ATTEMPT
Arkansas	224	19	11.8
LSU	311	26	12.0
Florida	296	23	12.9
Kentucky	368	27	13.6
Tennessee	345	23	15.0
South Carolina	323	20	16.2
Vanderbilt	337	16	21.1
Alabama	360	17	21.2
Auburn	257	12	21.4
Georgia	285	9	31.7
Ole Miss	257	8	32.1
Mississippi State	326	9	36.2

The Razorbacks' 19 touchdown passes have covered an average of 25 yards, including six between 35 and 56 yards.

"With a good running game, we want to be explosive in the passing game," Arkansas Coach Houston Nutt said. "When we do throw it, we're going for the end zone, and our guys believe we're going to score."

The Razorbacks lead the SEC and rank fourth nationally in rushing offense, averaging 230.6 yards per game.

Opposing defenses go into games focused on slowing down Arkansas sophomore tailbacks Darren McFadden and Felix Jones, and that leaves the receivers facing a lot of 1-on-1 coverage situations.

"The biggest part of our success in the passing game is our running game," Monk said. "Defenses get sucked in so much on our running game that we get single coverage, and when that happens, it's our job as receivers to make a play."

This will mark the fourth time in the past five seasons Arkansas has the SEC's best rushing offense, but that hasn't always translated into a big-play passing game. Last season, the Razorbacks averaged 216.9 rushing yards to lead the SEC and rank 12th nationally, but finished with just 13 touchdown passes in 11 games.

"They're a better team than they were last year by far," LSU senior strong safety Jessie Daniels said. "They do a lot more stuff."

A difference for Arkansas this season is more emphasis on the passing game with the addition of offensive coordinator and receivers coach Gus Malzahn and quarterbacks coach Alex Wood to the coaching staff.

Malzahn directed a Spread passing offense at Springdale High School to a 14-0 record last season, and Wood is a former NFL assistant coach.

"We're spending more time on the passing game and going out in the game and trying to make it happen," Monk said.

Improving the passing game was a focal point of spring practice.

"We've worked extremely hard on the pass, so you'd like to feel that hard work has paid off and made us better," Malzahn said. "Our quarterbacks and wide receivers have been patient and understand our team's strengths [is running], but they also understand when it's their time to make things happen."

Arkansas has connected on two touchdown passes in the final seconds of the first half, including a 50-yarder from sophomore Casey Dick to Monk at South Carolina and a 29-yarder from Dick to freshman Damian Williams at Mississippi State.

Monk also has touchdown catches for 56 yards at Vanderbilt and 50 at Auburn on throws from freshman quarterback Mitch Mustain.

"Overall, I just think we're more experienced and you're seeing that pay off," Dick said. "Guys know what they're supposed to do now."

McFadden also has given the passing game a boost with his play at quarterback out of the Shotgun formation the previous five games. He's completed all three of his pass attempts, including two touchdowns—9 yards to tight end Wes Murphy against Louisiana-Monroe and 12 to Monk against Tennessee.

"They are a big-play offense," LSU Coach Les Miles said. "They are the kind of offense that scores long touchdowns, and that style of formation [with McFadden taking direct snaps] is an opportunity for them.

"We're taking it seriously, and we're going to spend time on if to accomplish defending it well, and it's obviously more difficult in a short week."

Last week, Malzahn called a reverse pass—with McFadden taking the snap, handing off to Jones, who pitched to Dick, who threw to Monk—that resulted in a 35-yard touchdown play against Mississippi State.

Malzahn said the passing game has continued to improve each game.

"We haven't had to totally use it full force yet," Malzahn said. "But we've kept practicing it, and when the time comes, I feel we'll be able to do that."

Study Questions for Case Study 1.3

"UA passes produce a remarkable ratio" by Bob Holt
Arkansas Democrat-Gazette
November 22, 2006

1. Critique the table in the article. How are the data in the 3rd column obtained? Identify any errors in the table.

2. On LSU's next pass attempt, they throw for a touchdown. That is, the next pass attempt results in a touchdown. Find their new attempts/TD ratio.

3. Mississippi State is last in the conference with a passing touchdown every 36.2 attempts. Suppose that Mississippi State goes on a streak so that every pass attempt results in a touchdown. How many pass attempts would they need to lower their ratio and overtake Arkansas to be first in the conference in the attempts/TDs ratio?

4. Suppose instead that Mississippi State goes on a pace so that for every 3 pass attempts they make, they throw for one touchdown. How many attempts would Mississippi State need in order to overtake Arkansas for the conference lead in attempts/TDs ratio?

5. How many pass attempts would Mississippi State need to overtake Arkansas if their pace changed to a touchdown pass for every 5 pass attempts? Every 10 pass attempts? Every 25 pass attempts?

6. Suppose we state the question in more general terms. Assume Mississippi State maintains the pace of a touchdown for every n pass attempts. For what values of n will Mississippi State eventually overtake Arkansas?

7. Examine your solutions to questions 3–5. Identify your solution strategy for each question from the list below. Re-answer questions 3–5 using those strategies that you have not previously tried.

    a. Guess-and-check: I basically guessed at a possible answer and then checked to see if it was correct. If not, I modified my guess and tried again.

    b. Algebra: I set up an equation with an unknown and used algebra to solve for the unknown.

    c. Table: I created a table of values and used the table to identify the solution.

    d. Graph: I created a graph and used the graph to identify the solution.

Case Study 1.4: Harps Ad

Resource Material: Advertising poster used by Harps Food Stores.

Learning Goals: The learning goals of this case study include reviewing the ways quantities are compared and interpreting a common confusing way of comparing quantities.

Resource Material for Case Study 1.4

This case focuses on comparing quantities by illustrating a slogan that compares two quantities by the troublesome "times less."

Harps Food Stores, a grocer with stores in Arkansas, Missouri, and Oklahoma, used the slogan in the poster to indicate that the chicken they marketed had less sodium than that marketed by Harps' competitors.

1. **The poster states, "Our chicken contains up to 5 times less sodium [than the chicken of our competitor]." Explain carefully what this might mean.**

2. **Using your reasoning from Question 1, if Harps' competitor's chicken has 100 mg of sodium per serving, how much sodium does Harps' chicken have, assuming the ad is correct?**

3. **Explain what Harps probably meant by the phrase "five times less" and restate the** ad slogan with this interpretation. Discuss whether or not you think the new slogan is as effective as the one in the original ad.

4. **The dual statement to "times less" is "times more." Is 30 five times more than 6? Why or why not?**

Quantitative Reasoning

Section 2

Percent and Percent Change

In Section 2 we study percent, a technique for comparing two quantities, and percent change, a calculation that measures how much a quantity increases or decreases. The following items appear in this section.

- Introduction to percent and percent change.
- Case Study 2.1: Letters to the Editor on Tax Rates, Seven Letters to the Editor, *Arkansas Democrat-Gazette,* June and July, 2003.
- Case Study 2.2: "Other People's Money" by Paul Krugman, *New York Times,* November 14, 2001.
- Case Study 2.3: "Big Stink in Little Elkins" by Mike Masterson, *Arkansas Democrat-Gazette*, September 13, 2001.
- Case Study 2.4: "Study Finds That About 10 Percent of Young Males Are in Jail or Detention" by Sam Dillon, *New York Times*, October 8, 2009.
- Case Study 2.5: "Trainees fueling agency's optimism" by Charlotte Tubbs, *Arkansas Democrat-Gazette*, November 14, 2005.
- Case Study 2.6: "More Mothers of Babies Under 1 Are Staying Home" by Tamar Lewin, *New York Times,* October 19, 2001.

Introduction to Percent and Percent Change

Percent

Percent means "per 100," so 32% means "32 per 100" or "32 out of 100" or $\frac{32}{100}$.

"32% of 545 is 174.4" means $\frac{32}{100} \times 545 = 174.4$. More generally, "P% of N is A" means $\frac{P}{100} \times N = A$. The *base for a percent* is the quantity to which the percent applies. For example, in "35% of 700 is 245" the base for 35% is 700. Using the incorrect base for a percent is a very common error, and the base for commonly used percents vary and are often not stated explicitly.

For example, if you received 40 points on a 55-point quiz, you might interpret your score as $\frac{40}{55}$, or about 73%. Perhaps later in the course you received 20 points on a 25-point quiz for $\frac{20}{25} \times 100 = 80\%$, so your performance has improved from one quiz to the next. To compute the percentage you received on each quiz, you had to use a different base each time, 55 for the first quiz and 25 for the second quiz. Whenever you read about or compute a percent, give careful thought to determining the correct base.

Statements about percents appear in numerous news stories, media reports and informational brochures. They are simply *everywhere!* However, as we shall see, they can be used in many different ways. In order to understand the basics of percents, there are a few types of calculations one needs to master. As stated above, all these calculations are variations on the statement "P% of N is A," where one of these three quantities is unknown.

Example 1: Know P and N, find A. If a county proposes charging a tax of 0.5% (P) on a $270 (N) purchase, it is possible to determine the amount (A) of tax paid. One would need to find 0.5% of $270. We calculate:

$$\frac{0.5}{100} \times \$270 = \$1.35.$$

Example 2: Know P and A, find N. Suppose Jason transfers to a different school. If Jason has to drive 48 miles (A) to get to his new school and this represents 225% (P) of the distance he had to drive to get to last year's school, how far (N) did he have to drive last year? We calculate as follows:

$$\frac{225}{100} \times N = 48, \text{ so } N = \frac{48}{2.25}, \text{ or } N = \frac{4800}{225} = 21.3 \text{ miles.}$$

Example 3: Know N and A, find P. Suppose that 25 students enrolled in Math 101 last year and 75 enrolled this year. What percent (P) of 25 (N) is 75 (A)? We calculate:

$$\frac{P}{100} \times 25 = 75$$
$$\frac{P}{100} = \frac{75}{25} = 3$$
$$P = 300$$

The number of students enrolled this year is 300% of the number enrolled last year.

Example 4: Sometimes rates, say tax rates, are expressed in terms other than percent. For example, one mill is $1 per $1000, and taxes on property are often stated in terms of mills. Sometimes non-standard rates are given such as $6 per $1000. To change this to percent, one considers the following:

$$\frac{6}{1000} = \frac{x}{100}$$

where x will be the tax rate percent, or "rate per hundred." Note that we have converted from a rate that was per thousand to a rate that is per hundred, and the rate is $.60 per $100.

Example 5: Percent is just a special ratio where the denominator is 100. For example, 20% of 50 is a number x so that $\frac{20}{100} = \frac{x}{50}$ or $x = \frac{20}{100} \cdot 50 = (0.20) \cdot (50) = 10$.

Changes Measured by Percents

In many instances, it is important to state how much a particular quantity has increased or decreased. For example, if the population of a town was 1000 and it increased by 25%, the new population of the town would be 1250. Conversely, if a pair of shoes cost $60 and the price was reduced 30%, the new price would be $42.

In general, these situations can be represented as follows:

A is increased by P% to get B.

or

A is decreased by P% to get B.

In many of the problems in this section, you will be given any two of A, P, and B, and you will need to find the third.

For example, in the instances mentioned above, if the population of a town was 1000 and it increased by 25%, then we can see that 1000 is increased by 25%. This can be written as $1000 + \left(\frac{25}{100} \cdot 1000\right) = 1000 + 250 = 1250$. Likewise, if a pair of $60 shoes is marked down 30%, this can be written as $60 - \left(\frac{30}{100} \cdot 60\right) = 60 - 18 = 42$.

In general, these situations can be represented as follows:

A is increased by P% to get B means $A + \left(\frac{P}{100} \cdot A\right) = B$. This can be re-written as $A\left(1 + \frac{P}{100}\right) = B$.

A is decreased by P% to get B means $A - \left(\dfrac{P}{100} \cdot A\right) = B$. This can be re-written as $A\left(1 - \dfrac{P}{100}\right) = B$. A more general way to describe this situation is to consider one equation

$$A + \left(\dfrac{P}{100} \cdot A\right) = B$$

and adopt the convention that P can be either positive (when dealing with an increase) or negative (when dealing with a decrease).

Note that if one needs to emphasize the base for P%, one would write, "A is increased by P% of A to get B."

Example 6: *Know A and P, find B*. If 75000 is increased by 20%, the result is

$75000 + \left(\dfrac{20}{100} \times 75000\right) = 90000$ or $75000 \times (1 + 0.20) = 75000 \times 1.20 = 90000$.

Example 7: *Know A and P, find B*. If 75000 is decreased by 30%, the result is

$75000 - \left(\dfrac{30}{100} \times 75000\right) = 52500$ or $75000 \times (1 - 0.30) = 75000 \times .70 = 52500$.

Example 8: If 80 is increased by 20% and then the result is decreased by 25%, the final result is found in two steps. First, increase 80 by 20% to get $1.20 \times 80 = 96$. Second, decrease 96 by 25% to get $.75 \times 96 = 72$.

Now we focus on further examples of percent change where the two values are known but the percentage is not. This can be done by taking the previous equations and solving for P: $A\left(1 + \dfrac{P}{100}\right) = B \Rightarrow \dfrac{P}{100} = \dfrac{B}{A} - 1$ or $P = \dfrac{B - A}{A} \times 100$.

Example 9: Know A and B, find P. If the cost of a gallon of gasoline increases from $1.70 to $1.90 the percent increase is $\dfrac{\$1.90 - \$1.70}{\$1.70} \times 100 = 11.76\%$.

Example 10: If the cost of a gallon of gasoline decreases from $1.90 to $1.70 the percent change is $\dfrac{\$1.70 - \$1.90}{\$1.90} \times 100 = -10.53\%$. In general, negative percent changes are decreases and positive percent changes are increases.

Example 11: Know A and B, find P. By what percent P is 40 increased by to get 50? Another way to phrase this is to ask, 50 is what percent more than 40? Thus one calculates $P = \dfrac{50 - 40}{40} \times 100 = 25\%$.

In some instances, you will know the percent change P as well as the new value B, but you will not know the value A of which the percentage was taken (e.g., you will not know the base of the percent). The examples below illustrate these instances.

Example 12: If the population of a city in 2012 was 75,870 and this was an increase of 6% since 2002, what was the population in 2002?

To emphasize the base for the 6%, this can be stated as follows:

The population of a city in 2012 was 75,870. If A represents the population in 2002 and the 2012 population is A increased by 6% of A, find the 2002 population A. (Note: The base for the 6% is the quantity you do not yet know, the population in 2002.)

Note: The symbol P might be chosen for the unknown population, but we have already used P for the percent change. Therefore we choose A to represent the 2002 population.

Solution:

$$A\left(1 + \frac{P}{100}\right) = B$$
$$A\left(1 + \frac{P}{100}\right) = 75870$$
$$1.06A = 75870$$
$$A = \frac{75870}{1.06} = 71575$$

Example 13: If a laptop computer sells for $1100 in 2012 and this is a decrease of 7% in the price since 2010, what was the price in 2010?

Let A represent the price of the laptop computer in 2010. Then:

$$A\left(1 - \frac{P}{100}\right) = B$$
$$A\left(1 - \frac{7}{100}\right) = 1100$$
$$0.93A = 1100$$
$$A = \frac{1100}{0.93} = \$1183$$

Important Interlude on Terminology: To distinguish between the "increase in percent" and the "percent increase in percent," the phrase "percentage points" is used in stating the first and the phrase "relative increase" is used in the latter. Consequently, when a change in percentage is reported in the media, there is a chance of confusion. For example, if the percentage of students who drive cars on campus increases from 50% to 60% then "percent change" requires some thought. There is an increase in the "percent of students who drive cars on campus" from 50% to 60%, and that increase is 10%. That is the absolute change in the percentages. There is also the relative change that is $\frac{60\% - 50\%}{50\%} \times 100 = 20\%$. In order to distinguish the absolute change in percentage from the relative change in percentage, the absolute change is indicated as an increase of 10 percentage points.

This distinction is not always made in media reports. For example:

1. In a letter to the editor: "For the record, going from 1 percent proficient to 3 percent proficient is an increase of 200 percent." This is a relative increase.
2. In Case Study 2.5 this confusion is central. The graphic accompanying the article has a statement that the number of trainees increased by 15 percent, but the number of trainees increased by almost 30 percent. The increase of 15 percent is an absolute increase in the percent of trainee positions filled, which we would note as an increase of 15 percentage points.
3. Change in percentage points and relative change (percentage change in percentages) can send different messages. Below is an example taken from *Achieving Quantitative Literacy* by Lynn Arthur Steen (Mathematical Association of America, 2004, pp. 96–97). In this example the occurrence of strokes in women in a medical trial study increased from 0.2% to 0.28%, a percentage point increase of .08, which represented 8 more strokes among 10,000 women. The relative change is 40%.

Hormone Replacement Study

One of the major health stories of 2002 was the decision by NIH to end early a major clinical trial of hormone replacement therapy (HRT) due to increased risks of stroke and breast cancer. The day the HRT decision was announced, headlines, news articles, and the NIH press briefing itself all emphasized, for example, a 26% higher risk of breast cancer, a 29% increase in heart attacks, and a 41% higher risk of strokes among women taking the hormone therapy.

Virtually none of the early news accounts said what the actual risks were. What they did say, for example, is that of 10,000 women taking the combined therapy as compared with the same number taking a placebo, 8 more will have strokes each year. It takes a bit of algebra to discover that without HRT the risk of stroke is 20 per 10,000 (that is, 2 per thousand) and that HRT raises this to 28—a 40% increase. So the increase in annual risk of stroke is from .2% to just under .3%. These numbers convey quite a different emotional story than do the headlines of 41%.

Example 14: If 40% of Western University students read the campus newspaper, *Tattler*, in 1990, and 60% of Western students read *Tattler* in 2000, the increase in the percentage points of students who read *Tattler* is 60% – 40% = 20% and the percent increase in the percent of students (relative increase) who read *Tattler* is $\frac{60 - 40}{40} \times 100 = 50\%$.

Example 15: Suppose that there were 500 Central University students majoring in Journalism in 1990 and that in 2000 there were 700 Central University students majoring in Journalism.

a. The increase in students majoring in Journalism is 700 – 500 = 200.
b. The percent increase in students majoring in Journalism is $\frac{700 - 500}{500} \times 100 = 40\%$.

Example 16: Use the information in *Example 15* plus the following: In 1990 there were 14000 students enrolled at Central and in 2000 there were 15000 students enrolled at Central. Thus the percent of Central students who were Journalism majors in 1990 was $\frac{500}{14000} \times 100 = 3.75\%$ and in 2000 was $\frac{700}{15000} \times 100 = 4.67\%$, so

    a. The increase in the percent (increase in percentage points) of Central students majoring in Journalism was $4.67\% - 3.57\% = 1.1\%$. In this case our answer is 1.1 percentage points.

    b. The percent increase (relative increase) in the percent of Central students majoring in Journalism was $\frac{4.67 - 3.57}{3.57} \times 100 = 30.81\%$.

Percent and Percentage Word Clues

In normal usage, percent is used only when preceded by a number, as in 20 percent, or 20%. Otherwise, the word percentage is used to denote a ratio, as in the percentage of freshmen among students. Increasingly, though, percent is used as a stand-alone noun as above in the "percent of Central students." It is helpful in many cases of percentages less than 100% to identify the *whole* and the *part* in the ratio. For example, in considering the percentage of "students who are freshmen," the whole is students and the part is students who are freshmen. Earlier we called the *whole* the *base for the percent*. Frequently, words such as "of" and "among" are precursors of the whole and the part follows as a conditional "who" or "that."

There are uses of percent to compare quantities where the whole-part model does not reflect the actual situation. For example, if there are 40 male members and 60 female members of a band, then the number of female members is 150% of the number of male members. In this case the set of female members is not part of the set of male members. Of course, 40% of the band members are male and 60% of the band members are female, giving parts of male band members and female band members of the whole of band members.

Case Study 2.1: Letters to the Editor on Tax Rates

Resource Material: Seven Letters to the Editor, *Arkansas Democrat-Gazette*, June 30, 2003, to July 15, 2003.

Learning Goals: The learning goals of this case study include critical reading of source material and performing basic calculations with percents and ratios.

The initial letter of June 30 in this sequence of seven letters was responding to a previous letter that is not included. The initial letter is from Bob Massery. Six people wrote letters concerning Mr. Massery's statements regarding the tax rates of a person who pays $5,000 in tax on a $30,000 income and a person who pays $53,000 in tax on a $200,000 income. The six letters about Massery's letter, Pierce (July 9), Stilley (July 9), Basinger (July 10), McGuire (July 14), Herrington (July 15), and VanHook (July 15), believe Massery's calculation of tax rates are not correct. Some of the letter writers make errors when they attempt to correct Massery.

Warm Up Exercises for Case Study 2.1

1. Give the following tax rates in percent.

   a. $12.00 per $10000

   b. $3.75 per $225

   c. $7350 tax on an income of $47000

   d. $1 in taxes per $5.25 in income

   e. 42.5 mills

2. John pays $5000 taxes on an income of $35000 and Jane pays $12000 taxes on an income of $60000.

   a. Find and compare the tax rates of John and Jane in percents.

   b. Find the tax rates of John and Jane in dollars per $1000 income.

3. The school tax rate in a certain city is $3.50 per $1000 of assessed property value.

   a. Find this tax rate as a percent.

   b. Find this tax rate in the following form: A tax rate of $3.50 per $1000 of assessed property value is a tax rate of $1 per each $ _____ of assessed property value.

Articles for Case Study 2.1

Arkansas Democrat-Gazette—June 30, 2003

Taxpayer airs his woes

Re the letter that Roger Bresnahan of Hot Springs Village: I'm one of these people who make less than $30,000 a year. I pay almost $5,000 a year in federal income tax.

He says a person making $200,000 a year pays $53,000 in federal tax a year. So, my friend, a person [making] $30,000, pays $6 on $1,000: a person making $200,000 pays $3.77 on $1,000.

I wish I was one of those unlucky people paying $53,000 [so] I could also take my family out to eat every night and put it on an expense account. Oh, woe is me.

BOB MASSERY
Little Rock

Reprinted with permission from the *Arkansas Democrat-Gazette.*

Arkansas Democrat-Gazette—July 9, 2003

Taxes confuse everyone

Re Bob Massery's letter and his observation that he, paying almost $5,000 federal income tax on $30,000 income, paid $6 per $1,000 and someone else, paying $53,000 on $200,000 income, paid only $3.77 per $1,000: I think his numerators and denominators are mixed up. He paid $1 tax for every $6 income. The other person paid $1 tax for every $3.7 income.

Figuring taxes is enough to confuse any of us, regardless of income.

ANN PIERCE
Pine Bluff

Reprinted with permission from the *Arkansas Democrat-Gazette.*

Arkansas Democrat-Gazette—July 9, 2003

Math skills are lacking

How does letter writer Bob Massery survive on his math skills—or, more accurately, lack of same? Both of the computations in his recent letter were not only wrong, but woefully and obviously wrong. Arkansas public education is sadly deficient, but is it really this bad?

OSCAR STILLEY
Fort Smith

Reprinted with permission from the *Arkansas Democrat-Gazette.*

Arkansas Democrat-Gazette—July 10, 2003

Way off on his point

Re the letter from Bob Massery of Little Rock on his taxpayer woes: Who taught this man math?

He says since he make $30,000 a year and pays $5,000 in taxes, he pays $6 on $1,000, while a guy making $200,000 and paying $53,000 in taxes is paying only $3.77 on $1,000.

Let me redo the math for him. He is paying $166.67 per $1,000 in taxes and the rich guy is paying $265 per $1,000 in taxes. Not only is his math way off, he is way off on his point that the rich man pays a lower tax rate.

Depending on the size of Massery's family, he probably is getting most of the tax back as a refund—unless he's using the same style math on his tax return as he did in his letter.

PHILLIP BASINGER
Springdale

Reprinted with permission from the *Arkansas Democrat-Gazette.*

Arkansas Democrat-Gazette—July 14, 2003

Tax obligation revisited

Re the letter from Bob Massery saying that he paid almost $5,000 in taxes on $30,000, which is $6 per $1,000, and that a person making $200,000 will pay $53,000, which is $3.77 per $1,000: I couldn't figure out what he was talking about.

I figured the taxes on a person making $30,000 and a person making $200,000. They are both single and are using the standard deduction. The person making $30,000 will pay taxes on $27,000 and should pay $3,754 in taxes for 2002. That would be $125 per $1,000 income.

The person making $200,000 will pay taxes on $198,560 and should have paid $56,748 in 2002. That would be $284 per $1,000 of income.

DAN McGUIRE
Marshall

Arkansas Democrat-Gazette—July 15, 2003

Correcting tax figures

Re the letter from Bob Massery, "Taxpayer airs his woes": I am sure that Massery's heart is in the right place, but his math is not.

He claims that a person making $30,000 a year pays federal taxes of $6 on every $1,000 of income and that his tax bill is $5,000. He claims that a person making $200,000 a year pays federal taxes of $3.77 on every $1,000 of income and his tax bill is $53,000. His implication is that all of this is unfair.

Let's do the math correctly and see what's fair. I will use the numbers he provided in his letter. Five thousand dollars of taxes on $30,000 of income is not $6 per $1,000, it is $166.67 per $1,000. That represents 16.7 percent of that person's income.

Fifty-three thousand dollars of taxes on $200,000 is not $3.77 per $1,000, it is $265 per $1,000. That represents 26.5 percent of that person's income.

Fair or unfair? Depends on whom you ask. However, if we are to debate the relative merits of this or that tax code proposal, let's at least start with correct numbers.

BILLY HERRINGTON
Maumelle

Arkansas Democrat-Gazette—July 15, 2003

Using the funny math

I had to smile about the letter from Bob Massery of Little Rock. He says he pays $5,000 taxes on $30,000 income and calls this a $6-per-$1,000 rate. This is only .06 percent. Great rate.

But at this rate he should have paid only $150.

He also says a person making $200,00 and paying $53,000 in taxes is only paying $3.77 per $1,000. At this rate the person should have paid only $754. Actually, he is paying on the $30,000 16.67 percent, and the person making $200,000 is paying 26.5 percent.

How do these people come up with these numbers? This brings back memories of Al Gore's funny math. Please keep printing letters like this, and also Gene Lyon's column. They are both good for chuckles and remind us of what these people are really like.

EDSEL LEE VANHOOK
Hot Springs Village

Study Questions for Case Study 2.1

Seven Letters to the Editor
Arkansas Democrat-Gazette
June and July, 2003

1. Create an organized list for how the tax rates for $5,000 taxes on $30,000 income and $53,000 taxes on $200,000 income are stated in the seven letters under consideration.

2. Which of the stated rates are correct and which are incorrect? Be sure to support your conclusions with appropriate quantitative analysis.

3. Which, if any, of the letters dispute the amounts of tax cited by Mr. Massery: $5,000 on $30,000 and $53,000 on $200,000?

4. What is the mistake that Mr. Massery probably made in computing the tax rates?

5. Which of the six letters responding to Massery have errors, and what are those errors?

6. Which letter would you choose as the most appropriate rebuttal to Mr. Massery's letter? Write 2 or 3 sentences supporting your choice.

Case Study 2.2: Other People's Money

Resource Material: "Other People's Money" by Paul Krugman, *New York Times*, November 14, 2001.

Learning Goals: The learning goals of this case study include learning to calculate values quickly and accurately (compute on the fly), careful reading of text that includes quantitative arguments, understanding quantitative measures such as unemployment rates, and calculating with percentages.

This Op-Ed[3] piece offers several avenues of quantitative exploration and reasoning including number sense, percent and percent changes, understanding unemployment rates, reasonableness of answers, and rationales for different answers. There are several numbers, some very large, giving opportunities to develop understanding of the size of numbers and comparisons of various numbers.

Warm Up Exercises for Case Study 2.2

1. If you have $40 and this is 2% of the cost of a motorcycle you want to buy, find the cost of the motorcycle.

2. Two-fifths of one percent of the residents of Louisiana own 18,700 Toyota trucks.

   a. Find an approximation of the population of Louisiana.

   b. What assumption(s) are you making to get your approximation?

3. Assume the number of Eastern University students enrolled in the A & S College is 5,500 and the number of Eastern students enrolled in the Engineering College is 2,300.

   a. The Engineering enrollment is what percent of the A & S enrollment?

   b. The A & S enrollment is what percent more than the Engineering enrollment?

4. If the attendance at the first 2011 Eastern University football game was 76,500 and this attendance was an increase of 8% over attendance at the first football game in 2010, find the attendance at the first game in 2010.

---

3. Op-Ed (or oped) is a combination of opposite and editorial because such newspaper articles usually are opinion pieces and are placed on the page opposite the editorial page, which is where the newspaper publishes its opinions.

Article for Case Study 2.2

New York Times
November 14, 2001
Other People's Money
By PAUL KRUGMAN

You may have seen the story about the businessman who allegedly used the attack on the World Trade Center to make off with other people's money. According to his accusers, Andrei Koudachev stole $105 million that had been invested with his firm, falsely asserting that the sum had been lost in the collapse of the towers. It's not entirely clear whether he is accused of stealing the money before Sept. 11, then using the disaster to cover his tracks, or of taking the money after the fact; maybe both.

It's too bad that so many of our leaders are trying to pull the same trick.

Just before Sept. 11, political debate was dominated by the growing evidence that last spring's tax cut was not, in fact, consistent with George W. Bush's pledge not to raid the projected $2.7 trillion Social Security surplus. After the attack, everyone dropped the subject. At this point, it seems that nobody will complain as long as the budget as a whole doesn't go into persistent deficit.

But two months into the war on terrorism, we're starting to get a sense of how little this war will actually cost. And it's time to start asking some hard questions.

At the beginning of the week we learned that the war is currently costing around $1 billion per month. Oddly, this was reported as if it were a lot of money. But it's only about half of 1 percent of the federal budget. In monetary terms, not only doesn't this look like World War II, it looks trivial compared with the gulf war. No mystery there; how hard is it for a superpower to tip the balance in the civil war of a small, poor nation? At this rate, even five years of war on terrorism would cost only $60 billion.

True, the terrorist attack has also forced increased spending at home. But Mr. Bush has threatened to veto any spending on domestic security beyond the $40 billion already agreed. And even that sum is in doubt. Half of the $40 billion was money promised to New York; last week New York's Congressional delegation, Republicans and Democrats alike, demanded that Mr. Bush disburse the full sum, openly voicing doubt about whether he would honor his promise.

So the budgetary cost of the war on terrorism, abroad and at home, looks like fairly small change. Even counting the measures that are likely to pass despite Mr. Bush's threat, I have a hard time coming up with a total cost that exceeds $200 billion. Compare that with the $2.7 trillion Social Security surplus. What will happen to the remaining $2.5 trillion?

Again, no mystery: much of the money was actually gone before Sept. 11, swallowed by last spring's tax cut, which will in the end reduce revenue by around $1 trillion more than the numbers you usually hear. And the administration's allies in Congress are striving energetically to give away the rest in tax breaks for big corporations and wealthy individuals.

The new round of tax cuts is supposedly intended as post-terror economic stimulus. But recent remarks by Dick Armey give the game away. Defending the bill he and Tom DeLay rammed through the House—the one that gives huge retroactive tax cuts to big corporations—he asserted that it would create 170,000 jobs next year. That would add a whopping 0.13 percent to employment in this country. So thanks to Mr. Armey's efforts next year's unemployment rate might be 6.4 percent instead of 6.5. Aren't you thrilled?

Let's do the math here. This bill has a $100 billion price tag in its first year, more than $200 billion over three years. So even on Mr. Armey's self-justifying estimate, we're talking about giving at least $600,000 in corporate tax breaks for every job created. That's trickle-down economics without the trickle-down.

Ten weeks ago this bill, or the equally bad bill proposed by Senate Republicans, wouldn't have stood a chance. But now people who want to give the Social Security surplus to campaign donors think they can get away with it, because they can blame Osama bin Laden for future budget shortfalls.

They say every cloud has a silver lining. The dust cloud that rose when the towers fell has certainly helped politicians who don't want you to see what they're up to.

Study Questions for Case Study 2.2

"Other People's Money" by Paul Krugman
New York Times
November 14, 2001.

1. Choose one of the large dollar amounts in the article and describe the dollar amount in terms of something familiar enough to you so that you have a better understanding of the size of the dollar amount. For example, $600,000 buys 20 new automobiles.

2. From the information in the article, find the amount of the annual federal budget.

3. From the information in the article, find the number of people employed in this country. Be sure to explain your reasoning.

4. From the information in the article, find the number of unemployed in this country. Be sure to explain your reasoning.

5. Check the work after the sentence on the second page of the article, "Let's do the math here." Compare your results to those stated by the writer and explain any discrepancies.

Case Study 2.3: Big Stink in Little Elkins

Resource Material: "Big stink in little Elkins" by Mike Masterson, *Arkansas Democrat-Gazette*, Northwest Edition, September 13, 2001.

Learning Goals: The learning goals of this case study include recognizing the base of given percents, computing percent change, and distinguishing between absolute change in percent and relative change in percent.

This case study centers on percent changes in sewer charges where sewer charges are computed as some percentage of water charges. In the Elkins case sewer charges were to be changed from 100% of water charges to 85% of water charges. However, the intended drop by 15 percentage points was implemented as a drop of 11 percentage points. One lesson of the case involves determining what is meant by an "overcharge of 4%."

Warm Up Exercises for Case Study 2.3

1. A mathematical reasoning textbook sold for $40 in 2000. From 2000 to 2005, the price decreased 7%, and from 2005 to 2010 the price increased 20%. Find the price in 2010.

2. Assume the following: In 2005 there were 15,000 Central University (CU) students and 30% of them were freshmen, and in 2010 there were 17,000 CU students and 35% of them were freshmen.
   a. From 2005 to 2010, what is the (absolute) increase in the number of freshmen?
   b. From 2005 to 2010, what is the (absolute) increase in the percent of CU students who were freshmen?
   c. From 2005 to 2010, what is the percent (relative) increase in the percent of CU students who were freshmen?
   d. From 2005 to 2010, what is the percent (relative) increase in the number of freshmen?

3. If Jonesville's sewer charges are 70% of the water bills and this changes to 60% of the water bills, what is the percent reduction (relative) in sewer charges?

4. If a pair of shoes that costs $89 is on sale for $85, find the percent reduction.

5. If the price of a pair of shoes increases from $85 to $89, find the percent increase.

Article for Case Study 2.3

Arkansas Democrat-Gazette, **Northwest Edition**
September 13, 2001
Big Stink in Little Elkins
By MIKE MASTERSON

OK, so who put whatever in the Elkins water supply? Just as city fathers in this growing community southeast of Fayetteville flush one of the town's messes down the drain, its system invariably clogs with another odoriferous mess.

This time, it turns out that the bucolic city's administration, in mysterious ways no human can yet understand or explain, for years has been overcharging on its sewer rates and now must rebate $35,000 to nearly 600 customers. This discovery comes on the heels of the resignation, then reinstatement, of its police chief. In fact, those who have followed convolutions in Elkins law enforcement over the years have witnessed more twists and turns than a 1,000-mile Pig Trail.

But let's save the various police scandals for another day. This latest swirl is that somehow, Elkins sewer customers have wound up paying considerably more on their monthly bills than they should have. That is definitely raising the biggest stink in little Elkins nowadays.

The Elkins City Council voted in 1998 to reduce sewer rates by 15 percent. This rollback came a few years after the community's first sewer scandal, which, of course, is a story for another day.

Well, Elkins Mayor Oscar Lisle, who has made it clear that he was not the mayor back when the rollback was approved, said Elkins resident Will Pruett told him awhile back that somehow the 15 percent rollback had amounted to an 11 percent rollback. Pruett told Lisle that sewer bills in Elkins since 1998 were supposed to be running 85 percent of what the water bills are. But when Pruett put the pencil to what he had actually been paying, it amounted to 89 percent of the monthly water bill. His resulting call to Lisle led to this mystery of the Great Elkins Sewer Shortage.

Sewer users have been overpaying by 4 percent each month for three years. Lisle and others say that amounts to about $35,000 in overpayments by residential customers. They supposedly will be reimbursed in the form of credits on future sewer bills. The 100 former sewer customers who have since moved somehow will get checks—if city officials can locate them.

"We don't know why it happened," Lisle told reporter Matt Wagner. "But those rates were only rolled back by 11 percent. When I took office"—following the Elkins mayoral scandal, to be reviewed another day—"I never checked on what had been done here before.

We are going to have an accounting firm come in and do a thorough audit to see how much we owe in rebates."

The mayor said he had no idea how much the city may owe its commercial customers.

To compound this problem, the Elkins City Council is now considering yet another ordinance that would enable the city to raise sewer rates on current customers to repay what the city owes for its past blunder, and to comply with the state loan it obtained to build a water tower.

"We're not sure yet what to do," Lisle said. "We'll do whatever the accountants tell us to do."

The loan the city obtained to build a $200,000 water tower was based in part on the inflated sewer bills that Elkins has been wrongly collecting.

Audits are always a good idea when thousands of "miscalculated" dollars are at issue in any Arkansas community. But at this point, I'd like to offer yet another of my ideas in what is becoming a long list of unheeded ideas ranging from using the Bobby Hopper Tunnel as a high-speed vehicle wash to saving millions by installing $7 plastic seats in the remodeled Razorback Stadium.

Since the miscalculation of percentages has proven so costly here, I'd suggest in the future using any Elkins sixth-grade math class as a consultant to accurately determine the difference between 15 percent and 11 percent.

Generally speaking, there are 100 cents in every dollar, which means 15 percent would always equal 15 pennies. Applying that logic, 11 cents would always equal 11 percent. This fact would lead any 12-year-old to the inescapable conclusion that a basic difference of four pennies exists between 15 and 11 percent of every dollar. Fifteen would be the larger of the two.

It might also be wise to require Elkins city officials, including all the elected aldermen and women who pass such laws, to audit a sixth-grade math class for at least three months before assuming any position of significant public accountability.

Although they are adults, they would be permitted a recess along with the other students.

Mike Masterson is an award-winning Arkansas journalist. This article was published on page B5 of the Thursday, September 13, 2001 edition in the B5 section.

Study Questions for Case Study 2.3

"Big Stink in Little Elkins" by Mike Masterson
Arkansas Democrat-Gazette
September 13, 2001

1. Compare the two following statements from the article. Are they consistent? Your conclusion should be supported by careful and complete quantitative analysis.

    i) The Elkins City Council voted in 1998 to reduce sewer rates by 15%.

    ii) Pruett told Lisle that the sewer bills were supposed to be running 85% of what the water bills are.

2. Compare the following statements from the article. Are they consistent? Your conclusion should be supported by careful and complete quantitative analysis.

    i) But when Pruett put a pencil to what he had actually been paying, it amounted to 89 percent of the monthly water bill.

    ii) Sewer users have been overpaying by 4 percent each month for three years.

3. Use the information in the article to find how much the City of Elkins received from residential customers in total sewer payments during the three-year period following the 1998 reduction.

4. Consider a different version of the Elkins sewer cost reduction. Suppose that Elkins' sewer rates were 60% of one's water bill and the city decided to lower that to 45% of the water bill. Express this change in a reduction of percentage points and as a relative reduction.

Case Study 2.4: High Rate of Imprisonment Among Dropouts

> Resource Material: "Study Finds High Rate of Imprisonment Among Dropouts" by Sam Dillon, *New York Times*, October 8, 2009.

Learning Goals: The learning goals of this case study include identifying the parts-whole relationships of percentages presented in the media.

Article for Case Study 2.4

The New York Times
October 8, 2009
Study Finds High Rate of Imprisonment Among Dropouts
By SAM DILLON

On any given day, about one in every 10 young male high school dropouts is in jail or juvenile detention, compared with one in 35 young male high school graduates, according to a new study of the effects of dropping out of school in an America where demand for low-skill workers is plunging.

The picture is even bleaker for African-Americans, with nearly one in four young black male dropouts incarcerated or otherwise institutionalized on an average day, the study said. That compares with about one in 14 young, male, white, Asian or Hispanic dropouts.

Dropouts in Jail or Detention

Male high school dropouts were 47 times more likely than a college graduate to be jailed. And black men who dropped out had a much higher chance of incarceration than men in other groups, according to a study by the Northeastern University in Boston.

Males ages 16 to 24 who were incarcerated in 2006-7

High school dropouts	9.4%
High school students	1.5
High school graduates	2.8
1 to 3 years of college	1.1
College students	0.4
B.A. degree or higher	0.2

Male high school dropouts ages 16 to 24 who were incarcerated in 2006-7

Black	22.9%
Asian	7.2
White	6.6
Hispanic	6.1

The New York Times

Researchers at Northeastern University used census and other government data to carry out the study, which tracks the employment, workplace, parenting and criminal justice experiences of young high school dropouts.

"We're trying to show what it means to be a dropout in the 21st century United States," said Andrew Sum, director of the Center for Labor Market Studies at Northeastern, who headed a team of researchers that prepared the report. "It's one of the country's costliest problems. The unemployment, the incarceration rates—it's scary."

A coalition of civil rights and public education advocacy groups and a network of alternative schools in Chicago commissioned the report as part of a push for new educational opportunities for the nation's 6.2 million high school dropouts.

"The dropout rate is driving the nation's increasing prison population, and it's a drag on America's economic competitiveness," said Marc H. Morial, the former New Orleans mayor who is president of the National Urban League, one of the groups in the coalition that commissioned the report. "This report makes it clear that every American pays a cost when a young person leaves school without a diploma."

The report puts the collective cost to the nation over the working life of each high school dropout at $292,000. Mr. Sum said that figure took into account lost tax revenues, since dropouts earn less and therefore pay less in taxes than high school graduates. It also includes the costs of providing food stamps and other aid to dropouts and of incarcerating those who turn to crime.

Daniel J. Losen, a senior associate at the Civil Rights Project at the University of California, Los Angeles, said the study was consistent with other economic studies of the dropout crisis, though he said the methodology of its cost-benefit analysis "lacked transparency."

"The report's strength is that it reveals in clear terms that there's a real crisis with the high numbers of young, especially minority males, who drop out of school and wind up incarcerated," Mr. Losen said.

Previous studies have come up with estimates of the same order of magnitude on the social cost of low graduation rates. A 2007 study by Teachers College, Princeton and City University of New York researchers, for instance, estimated that society could save $209,000 in prison and other costs for every potential dropout who could be helped to complete high school.

The new report, in its analysis of 2008 unemployment rates, found that 54 percent of dropouts ages 16 to 24 were jobless, compared with 32 percent for high school graduates of the same age, and 13 percent for those with a college degree.

Again, the statistics were worse for young African-American dropouts, whose unemployment rate last year was 69 percent, compared with 54 percent for whites and 47 percent for Hispanics. The unemployment rate among young Hispanics was lower, the report said, because included in that category were many illegal immigrants, who compete successfully for jobs with native-born youths.

The unemployment rates cited for all groups have climbed several points in 2009 because of the recession, Mr. Sum said.

Young female dropouts were nine times more likely to have become single mothers than young women who went on to earn college degrees, the report said, citing census data for 2006 and 2007.

The number of unmarried young women having children has increased sharply in some communities in part, Mr. Sum said, because large numbers of young men have dropped out of school and are jobless year round. As a result, young women do not view them as having the wherewithal to support a family.

"None of these guys can afford to own a home, they just don't have any money," he said. "And as a result, any time they father a child it's out of wedlock. It wasn't like this 30 years ago."

He cited his hometown, Gary, Ind., as an example. "Back in the 1970s, my friends in Gary would quit school in senior year and go to work at U.S. Steel and make a good living, and young guys in Michigan would go to work in an auto plant," he said. "You just can't do that anymore. Today, you have a lot of dropouts who are jobless year round."

Study Questions for Case Study 2.4

"Study Finds High Rate of Imprisonment Among Dropouts"
Sam Dillon
New York Times
October 8, 2009

1. If three out of every 100 females with college degrees become single mothers, what percent of female high school dropouts become single mothers? Explain your reasoning and cite the information in the article that you used to answer this question.

2. What are the bases (wholes) and parts for the percentages of dropouts, high school graduates, and college degree holders that were unemployed in 2008? Explain each of the three, writing a sentence that clearly indicates the whole-part relationship.

3. What are the bases (wholes) and parts of the unemployment rates for young African-American dropouts, white dropouts, and Hispanic dropouts? Explain each of the three, writing a sentence that makes clear the whole-part relationship.

4. Use the information in the article to explain the statement that "male high school dropouts were 47 times more likely than college graduates to be jailed." Illustrate this with statements of relative frequency. That is, "approximately ___ of 1000 high school dropouts were in jail in 2006–7, compared to approximately ___ of 1000 college graduates."

5. What are the wholes and parts of the 9.4% and the 22.9% in the tables entitled "Dropouts in Jail or Detention"? Explain these by writing sentences that clearly indicate the whole-part relationships.

6. The sum of the percentages in the table entitled "Males ages 16 to 24 who were incarcerated in 2006–7" does not equal 100%. Provide a reasonable explanation for this observation.

Case Study 2.5: Trainees Fueling Agency's Optimism

Resource Material: "Trainees fueling agency's optimism" by Charlotte Tubbs, *Arkansas Democrat-Gazette*, November 14, 2005.

Learning Goals: The learning goals of this case study include identifying the base for a percent and distinguishing between an absolute increase in percents and relative percent increase in a quantity.

This case study provides revealing instances of the importance of precise language when discussing change and percent rate of change. The article centers on an increase in the percent of available positions filled. The State of Arkansas authorizes State agencies such as the one in this article, Division of Children and Family Services (DCFS), to employ a certain number of persons in positions. The positions here are caseworker trainees. According to the graphic, DCFS was authorized to employ 141 trainees in August, 143 in September, and 147 in October of 2005. In these three months, the numbers of trainee positions filled were 91, 108, and 118. The percent of positions filled in these three months all have different bases, and the change in these percents is confused with the percent changes in the number of trainees.

Warm Up Exercises for Case Study 2.5

1. In 1990 Central College's marching band had 50 women and 40 men as members; in 2010 the Central band had 60 women and 50 men.

   a. Find the percent of women in the band in 1990 and 2010.

   b. Find the change in percent of women in the band from 1990 to 2010.

   c. Find the change in the number of band members from 1990 to 2010.

   d. Find the percent change in the size of the band from 1990 to 2010.

   e. Find the percent change in the number of women band members from 1990 to 2010.

   f. Find the percent change in the percent of women band members (relative change) from 1990 to 2010.

Article for Case Study 2.5

Arkansas Democrat-Gazette
November 14, 2005
Trainees Fueling Agency's Optimism
State looks to fill child welfare spots
By CHARLOTTE TUBBS

The state's child welfare director is hailing an increase in caseworker trainees as a hopeful sign that an agency is making progress toward filling vacant positions.

The Division of Children and Family Services of the Department of Health and Human Services has struggled to keep caseworker positions filled since budget cuts and a hiring freeze that lasted through parts of 2003–04.

Caseworker trainees increasing

The number of Division of Children and Family Services caseworker trainees increased by 15 percent from August to October. Observers say this increase is a hopeful sign for the agency, which has eperienced a shortage of caseworkers.

CASEWORKER TRAINEES
Trainees
KEY
Total trainee positions

	Aug.	Sept.	Oct.
Total	141	143	147
Trainees	91 (65%)	108 (76%)	118 (80%)

More vacant positions increased the caseload of remaining caseworkers, increasing the turnover rate.

"Are we there yet?" said division Director Roy Kindle. "No, but I think we are turning the corner to get there."

Two members of a group working with child welfare administrators toward improvements at the division welcomed the report of more trainees, but said much work remains to solve the agency's problems.

Although an increasing number of caseworker recruits is a hopeful sign for the future of child welfare services, the upswing does not offer an immediate solution since training takes six months.

Separate reports from three organizations—the division, the non-partisan and nonprofit Arkansas Advocates for Children and Families, and an Arkansas Supreme Court committee—blamed the lack of caseworkers for delaying services such as counseling, plans for placing foster children in permanent homes and investigations into reports of neglect and abuse.

As of Aug. 31, 65 percent of trainee positions—or 91 out of 141—were filled, which increased to 80 percent of filled trainee positions—118 out of 147—as of Oct. 31.

Vacancy rates for other categories of child welfare workers, including caseworkers, caseworker specialists and supervisors, remained about the same during the three months.

Kindle said that the agency must not only recruit and retain workers, but also improve supervisors' skills with the hopes of providing more support to caseworkers.

"If [caseworkers] don't have a support structure readily available, it will kill us every time," Kindle said.

Salaries for caseworkers start at about $25,400, although in July a $2,500 increase in the starting salaries for four northwest counties took effect.

"At some point down the line, we will have to look at salaries, period," Kindle said, adding that staffing numbers will also require review.

Kindle credited the increase in caseworker trainees to new recruiting efforts by the division, including a co-op program and paid internships to give college students working experience before they graduate.

The division has also expanded its stipend program, which pays a portion of a student's education expenses in exchange for the student agreeing to work for the division for a set time period.

Division recruiters are also planning to conduct interviews at college campuses, something that has not been done in the past.

Several dozen trainees will be joining the ranks of caseworkers soon.

Two classes, which average 12 to 16 trainees each, graduated in Little Rock on Nov. 2.

A class in Jonesboro is scheduled to graduate Nov. 30, and another class in Arkadelphia is scheduled to graduate Dec. 14.

Consevella James, a member of a group working with the division, said the increasing trainee numbers are encouraging.

James is executive director of Treatment Homes Inc., which contracts with the division to place foster children with behavioral or psychiatric problems in therapeutic foster homes, and has seen the effects of caseworker vacancies firsthand.

"What has happened since we've had the shortage is the turnover has been so significant that a child may have five or six caseworkers over a period of six months," James said.

"That really does not help facilitate permanency for a child."

The increasing number of caseworker trainees shows potential for more caseworkers, but is no guarantee, James said.

"There is also the issue of competition with other private corporations and private businesses," she said.

Only when caseworkers leave training do they experience a full caseload, which currently averages about twice the recommended amount, and realize the emotional and time-consuming demands of the job, she said.

"For the salary that they are being paid, I think many individuals are not willing to invest that much time," James said.

Rich Huddleston, also a member of the group working with the division and executive director of Arkansas Advocates for Children and Families, said the increase in trainees represents progress.

But the newest trainees won't become caseworkers for another six months.

"I think you still have to be concerned for the kids who are in the system now and for the next six to nine months," Huddleston said. "For the next six months, we need to be carefully monitoring what's happening to kids out there."

He said he was encouraged by a report from the division that caseworkers are receiving more support equipment such as cell phones, laptops and car seats. The equipment should help ease the workload for existing caseworkers, he said.

James added that news of increasing caseworkers left her feeling optimistic, but she knows change will take time.

"I guess the question will be: How much time do we have when we're dealing with children's lives?"

This article was published on pages 1B and 9B of the Monday, November 14, 2005 edition in the Northwest Arkansas section.

Study Questions for Case Study 2.5

"Trainees Fueling Agency's Optimism" by Charlotte Tubbs
Arkansas Democrat-Gazette
November 14, 2005

1. Using the data in the article, find the percent increase in the number of trainees from August 31 to September 30 and from September 30 to October 31.

2. Explain carefully what is meant by the percents 65%, 76%, and 80% that are written on the bars in the graphic. Your explanation should include a brief description of how the percents were calculated as well as a verbal description of what the percents mean.

3. The text that accompanies the graphic includes: "Caseworker trainees increasing. The number of Division of Children and Family Services caseworker trainees increased by 15% from August to October." Do the given data support these two statements? If so, explain carefully why these statements are correct. If not, give the correct conclusions that are supported by the data provided. All explanations should be supported with relevant quantitative analysis.

4. Other increases that are relevant in discussing the topic of this article are the increases in the total positions, whether filled or not. Find these three percent increases.

5. Critique the graphic in the article. Is the graph effective in conveying the information? If so, tell why, and if not, describe a graphic that would convey the information better.

Case Study 2.6: More Mothers of Babies Under 1 Are Staying Home

Resource Material: "More Mothers of Babies Under 1 Are Staying Home" by Tamar Lewin, *New York Times*, October 19, 2001.

Learning Goals: The learning goals of this case study include careful reading of compound descriptions of portions and bases for percentages and comparing and contrasting different percentages that sound similar.

This is a study of confused bases for percents. For examples, consider the following pairs of phrases:

A. The percentage of men who are overweight
B. The percentage of men among overweight people

C. The percentage of teenagers who smoke and attend college
D. The percentage of teenage college students who smoke

Phrases A and B differ. To compute the percentage in A, one divides the number of overweight men by the number of men and multiplies by 100. To compute the percentage in B, one divides the number of overweight men by the total number of overweight people and multiplies by 100. More specifically, consider a set of 500 people that contains 250 men, 250 women, 200 overweight people, and 120 overweight men. The percentage of men who are overweight is $\frac{120}{250} \cdot 100 = 48\%$. The percentage of men among overweight people is $\frac{120}{200} \cdot 100 = 60\%$.

Phrases C and D differ also. Let's compute these two percentages using the following numbers: 2000 out of 5000 teenagers attend college; 500 of the 5000 smoke, and 200 of these attend college. The percentage of teenagers who smoke and attend college is $\frac{200}{5000} \cdot 100 = 4\%$. The percentage of teenage college students who smoke is $\frac{200}{2000} \cdot 100 = 10\%$.

Warm Up Exercises for Case 2.6

Alpha College has 10,000 students. Of these students, 4,000 live on campus and 6,000 live off campus. Of the 4,000 students who live on campus, only 1,000 have part-time jobs; and only 4,000 of the students who live off campus have part-time jobs.

1. Construct a diagram or a table that sorts the number of Alpha College students first by living on or off campus and then by part-time job or no part-time job.
2. Find the percent of Alpha College students who live on campus.
3. Find the percent of Alpha College students who live off campus.
4. Find the percent of Alpha College students who have part-time jobs.
5. Find the percent of Alpha College students living off campus who have part-time jobs.
6. Find the percent of Alpha College students having part-time jobs who live off campus.

Article for Case Study 2.6

New York Times
October 19, 2001
More Mothers of Babies Under 1 Are Staying Home
By TAMAR LEWIN

After a quarter-century in which women with young children poured into the workplace, last year brought the first decline in the percentage of women who have babies younger than 1 year old and are in the work force.

A new Census Bureau report said 55 percent of women with infants were in the labor force in June 2000, compared with 59 percent two years earlier. It was the first decline since 1976, when the government began tracking the numbers.

"In the late 1990's, the economy was going very strong," said Martin O'Connell, an author of the report. "I see two principles operating here, obvious ones that I've even heard expressed by a couple of pregnant women in the Census Bureau. One is that as more women delay their childbearing, which many now do until their 30's or 40's, they have a chance to build up a nest egg that allows them to take more time off. And the other is that, looking down the road in a good economy, there's the anticipation that it will be easy to find a job whenever you want to go back to work."

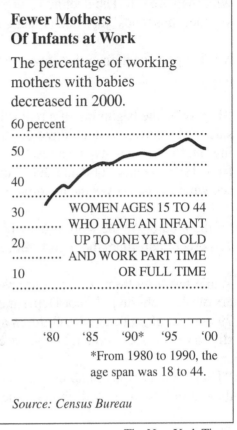

Fewer Mothers Of Infants at Work

The percentage of working mothers with babies decreased in 2000.

WOMEN AGES 15 TO 44 WHO HAVE AN INFANT UP TO ONE YEAR OLD AND WORK PART TIME OR FULL TIME

'80   '85   '90*   '95   '00

*From 1980 to 1990, the age span was 18 to 44.

Source: Census Bureau

The New York Times

The recent decline comes as no surprise to many economists, who say that while the number of mothers in the work force grew rapidly from the 1970's through the early 1990's, it stabilized in the late 1990's while there was a stream of publicity about successful women who quit work after having babies.

The Census Bureau report, "Fertility of American Women, 2000," said the most pronounced decline in working mothers was among married women, to 54 percent last year—the 1994 level—from 60 percent in 1998. White mothers, and those who had at least a year of college were also less likely to be working last year than in 1998. But there was no such decline among mothers who were single, African-American or Hispanic or did not pursue education beyond high school.

"I don't know where it goes from here, given the obvious fact that we're in something of a downturn, where jobs are going to be harder to find," said Howard Hayghe, a labor economist at the Bureau of Labor Statistics. "The hypothesis, if you will, is that moms with kids under a year old are leaving the labor force because they feel more comfortable with their wealth. The decline is largely among wives. So what's going to happen if they see their husbands' jobs and careers threatened?"

Most experts cautioned that it was far too early to regard last year's decline as the start of a trend.

"It may be the beginning of a real shift in what mothers of young children choose to do," said Vicky Lovell, a study director at the Institute for Women's Policy Research in Washington, "but it may just be a response to the unusual economic environment of the period from 1998 to early 2000 and that economy isn't there anymore. It will be interesting to see what happens over the next two years."

Of the mothers in the work force who had infants, 34 percent worked full time, and 17 percent part time, the report said. Four percent were unemployed but wanted to work.

While fewer mothers of babies are working, there has been no such decline among mothers of older children. Labor Department statistics, gathered yearly in March, indicated 79 percent of mothers with children ages 6 to 17 were in the work force in 2000—up slightly from 78.4 percent two years earlier.

"There's more concern about child care available to very young children, and once they're in school, it's a different issue for parents," Ms. Lovell said.

Though most mothers are now in the work force, several polls in recent years have found that most Americans believe that women with children—especially young children—should stay home and rear them.

Until recently, mothers who had both an infant and at least one older child were far less likely to be in the work force than were those who had only an infant. As recently as 1995, 59 percent of the women with only an infant were working, compared with 49 percent of those who had other children as well.

But that difference has been disappearing in recent years. In the 2000 report, the Census Bureau found little difference in the percentage of working mothers with only a baby and those with at least one older child as well.

Study Questions for Case Study 2.6

"More Mothers of Babies Under 1 Are Staying Home" by Tamar Lewin
New York Times
October 19, 2001

1. **Locate in the article (writing down where each is found) the following assertions that are parts of the main quantitative message.**

   Note: In the following, infant is used as synonymous with baby under 1.

   I.   More mothers of babies under 1 are staying home.

   II. After a quarter-century in which women with young children poured into the workplace, last year brought the first decline in the percentage of women who have babies younger than 1 year old and are in the work force.

   III. A new Census Bureau report said 55 percent of women with infants were in the labor force in June 2000, compared with 59 percent two years earlier.

   IV. Fewer mothers of infants at work

   V.   The percentage of working mothers with babies decreased in 2000.

   VI. Of the mothers in the work force who had infants under 1, 34 percent worked full time, and 17 percent part time, the report said. Four percent were unemployed but wanted to work.

2. **What are the relationships of the six assertions? Which assertions say essentially the same thing? Which assertions differ? Be sure to explain your reasoning by paying careful attention to the type of quantity being described in each statement.**

3. **Do you find other assertions that are part of the main quantitative message? If so what are they and how do they relate to the six listed above?**

4. **The first paragraph of the online version of this article is different from the print version's first paragraph. The online version read (II above):** *After a quarter-century in which women with young children poured into the workplace, last year brought the first decline in the percentage of women who have babies younger than 1 year old and are in the work force.* **The print version is seen below. Compare and contrast these two first paragraphs and describe carefully what percent is being described in each paragraph.**

5. **Which one of the above six assertions appears to be the correct message of the article?**

6. **Do assertions I and IV follow from assertion III? Support your conclusion with relevant and correct quantitative analysis.**

> By TAMAR LEWIN
> After a quarter-century in which women with young children poured into the workplace, the percentage of women in the labor force who had babies younger than 1-year old declined last year.

7. **Does assertion II follow from assertion III? Support your conclusion with relevant and correct quantitative analysis.**

8. **In assertion VI, what is the base of the 34 percent, 17 percent and 4 percent?**

9. **How should assertion VI be stated so that it is correct?**

Quantitative Reasoning

Section 3

Measurement and Indices

In Section 3 we study measurement and units. In particular, we focus on measurements of central tendency and the use of indices (or indexes) and units to compare quantities. The content of Section 3 is listed below.

- Case Study 3.1: "Forbes hospital rates high in heart care" by Luis Fabregas, *Pittsburgh Tribune-Review*, June 29, 2007.
- Case Study 3.2: "This year, Oscar smiles on familiar films" by Scott Bowles, *USA Today*, February 22, 2011.
- Case Study 3.3: "Real Estate Track," *The Morning News of Northwest Arkansas,* March 1, 2009, and "Median home prices fell nationwide in fourth quarter: NAR report," *The Seattle Times*, February 12, 2009.
- Introduction to Indices.
- Case Study 3.4: "Tell the Truth: Does this Index Make Me Look Fat" by Gina Kolata, *New York Times,* November 28, 2004.
- Case Study 3.5: "Market Gauges," *New York Times,* September 22, 2007.
- Case Study 3.6: Chuck Taylor All Star Advertisement.
- Case Study 3.7: "GOP Disputes," Associated Press Graphic, March 4, 2003, and "Is the glass half full . . . ," *Arkansas Democrat-Gazette,* May 18, 2003.
- Case Study 3.8: "*FOXTROT* cartoon" by Bill Amend, July 21, 2004.

Case Study 3.1: Forbes Hospital Rating

Resource Material: "Forbes hospital rates high in heart care" by Luis Fabregas, *Pittsburgh Tribune-Review*, June 29, 2007.

Learning Goals: The learning goals of this case study include understanding the definitions of mean and median of a set of numerical data and the interpretation of the term average.

This article reports on a ranking of hospitals based on death rates of patients admitted for either heart attack or heart failure. Over 4,000 hospitals are considered in the report and only a few receive above average ratings, and only a few receive below average ratings, leaving most of the hospitals in the average category. Aside from understanding how the ratings are most likely determined, there are language issues to be considered. For example, the headline has more than one possible interpretation. Another language issue is interpreting phrases such as "better-than-average death rate."

Warm Up Exercise for Case Study 3.1

1. Find the mean and median of the following sets of numbers:

    a.   191, 45, 68, 83, 64, 72, 54.

    b.   75, 18, 43, 34, 81, 56, 48, 65

2. How was computing the median different in parts (a) and (b)? How many of the numbers in each set are larger than the mean, and how many are smaller? How many of the numbers in each set are larger than the median, and how many are smaller?

3. Devise a set of ten numbers so that nine of them are "above average;" that is, nine of the numbers are larger than the mean. Is it possible to have all of them "above average," like Lake Wobegon's children? Explain a scenario for how this could happen, and explain a scenario when this is impossible. (Lake Wobegon is the fictional town in Minnesota that Garrison Keillor of Minnesota [and National] Public Radio's A Prairie Home Companion claims as his hometown.)

4. Devise a set of ten numbers such that eight of them are "average," one is above average, and one is below average.

5. Create a data set of ten test scores such that the mean is 80 and the median is 85.

6. In a certain class, grades are determined by semester percentages as follows: an A is 90 to 100 percent; a B is 80 up to 90 percent; a C is 70 up to 80 percent; a D is 60 up to 70 percent; and an F is below 60 percent. If C is the grade for average, then give the percentage intervals for ABOVE AVERAGE, AVERAGE, and BELOW AVERAGE.

Article for Case Study 3.1

Tribune-Review
June 29, 2007
Forbes hospital rates high in heart care
A Web site reports that 38 of 4,000 facilities in the United States are better than average
By LUIS FABREGAS

A first-of-its-kind Web site that compares death rates for several heart conditions nationwide is getting mixed reviews from local hospitals and consumer advocates, though most agree it should pressure hospitals into improving patient care.

"It raises the bar," said Dr. Michael L. Steinfeld, director of telemetry and noninvasive cardiology at West Penn Hospital-Forbes Regional Campus. "We don't do everything perfectly, so there's a lot of room for improvement in all kinds of care."

As it turns out, Forbes is helping set higher standards.

The Monroeville hospital was the only one in Pennsylvania to have a better-than-average death rate for heart failure patients. Only 38 of more than 4,000 hospitals nationwide were better than average.

> *"The best (the Web site) will do is steer you away from the exceptionally bad places"*
>
> ARTHUR LEVIN DIRECTOR OF CENTER FOR MEDICAL CONSUMERS, SAYING THE WEB SITE SHOULD PROVIDE ACTUAL DEATH RATES FOR INDIVIDUAL HOSPITALS

On average, 18 percent of patients nationwide died within 30 days of being admitted to a hospital for a heart attack. For heart failure, the death rate is about 11 percent. Most hospitals fell within those averages.

Officials at Forbes, one of six hospitals operated by West Penn Allegheny Health System, said they are proud of their performance.

"These results demonstrate that Forbes is a quality provider for heart failure patients," said Darlette Tice, the hospital's vice president of operations and chief nurse executive, noting that doctors at Forbes treat more than 700 heart failure patients every year.

Although death rates for some common procedures and treatments in Pennsylvania hospitals have been analyzed and published by the Pennsylvania Health Care Cost Containment Council, the new Web site for the first time compares heart attack and heart failure rates from 4,453 hospitals in the United States.

But the site does not provide actual death rates for individual hospitals. Arthur Levin, director of the New York-based Center for Medical Consumers, said that makes the information less valuable.

Reprinted from the *Pittsburgh Tribune-Review,* June 29, 2007.

"Is it a good thing? Yes. Is it useful? No," Levin said. "The best it will do is steer you away from the exceptionally bad places. So I guess it's a beginning, but a very limited beginning."

The Web site also gives information on how often hospitals provide key treatments known to get the best results for heart attack and heart failure patients.

Forbes officials said their employees have worked to improve those treatments for nearly a decade.

That includes everything from prescribing blood pressure-lowering drugs called ACE inhibitors to providing smoking cessation counseling.

"All those individual measures have played out ultimately in cutting mortality," Steinfeld said.

At the University of Pittsburgh Medical Center, officials said the measurement of such treatments at a national level has prompted them to organize quality groups to make sure the standards are met.

One such treatment for heart attack patients, for example, is the opening of a blocked artery with a small balloon within 90 minutes of the patient's arrival at the hospital.

"Our average just fell like a rock, as a result of a lot of changes," said David Sharbaugh, senior director of UPMC's center for quality improvement and innovation.

Sharbaugh said changes in treatment are more telling about a hospital's quality than actual death rates.

"You can't look at heart failure mortality and say, 'I'm going to find a silver bullet,'" Sharbaugh said. "You can't imply that if you're having a high mortality rate, they're killing people. People with heart failure are going to die. It's a matter of managing it."

Study Questions for Case Study 3.1

"Forbes hospital rates high in heart care" by Luis Fabregas
Pittsburgh Tribune-Review
June 29, 2007

1. What are some possible interpretations of the headline, *Forbes hospital rates high in heart care*, to this article? Is the report positive or negative about the Forbes hospital?

2. What journalistic practice would help you read the headline correctly?

3. The article's sub-head, *A Web site reports that 38 of 4,000 facilities in the United States are better than average,* indicates that very few of the hospitals rated are better than average.

   a. Is it possible for only 38 out of 4,000 numbers to be above average? Explain your reasoning.

   b. Give a real-world example of a distribution of scores in some population that would have an analogous number of scores "better than average." Be sure to explain why the distribution you chose meets this criterion.

4. The article contains the phrase "better-than-average death rate." How could this be interpreted? What does it mean in the article?

5. Use the information in the article to estimate the number of heart failure patients at the Forbes hospital that die each year. Explain your reasoning.

6. Analyze the final sentence of the 5th paragraph, which reads, "Most hospitals fell within those averages." What does this mean? For example, does it mean that most hospitals had a death rate of 11% within 30 days for patients admitted for heart failure? If it does not mean this, give a possible meaning.

Case Study 3.2: Best Picture Nominees

Resource Materials: "This year, Oscar smiles on familiar films" by Scott Bowles, *USA TODAY*, February 22, 2011.

Learning Goals: The learning goals of this case study include understanding the differences between mean and median and using that information to make inferences about data that are not given.

Warm Up Exercises for Case Study 3.2

1. Eight of ten people have weights of 129 pounds, 130 pounds, 140 pounds, 160 pounds, 170 pounds, 125 pounds, 220 pounds and 250 pounds. If the average (mean) weight of the ten people is 145 pounds, find the median weight.

2. If the median of four positive numbers is 20, the largest number is 50 and the next largest one is 30, find the other two numbers.

This year, Oscar smiles on familiar films

Nominees score big at box office
By Scott Bowles
USA TODAY

LOS ANGELES—Oscar is no stranger to heralding big films, but you may notice something you're not accustomed to in Sunday's Academy Awards telecast: movies you've actually seen.

Last week, the ballet drama *Black Swan* crossed the $100 million mark, bringing this year's crop of blockbuster best-picture nominees to five, or half the slate. While *Toy Story 3* ($415 million) and *Inception* ($293 million) were summer smashes, films such as *Swan, The King's Speech* and *True Grit* not only became unexpected hits but also awards-season favorites.

And with *The Social Network* just off the mark with $97 million and the boxing drama *The Fighter* ringing up $88 million, the Oscars are looking a little like the People's Choice Awards.

"It should help the Oscars in that a lot of people have seen the strongest contenders," says Brandon Gray, president of Box Office Mojo, which tabulates Academy Award contenders' ticket sales. "It adds a rooting element to the telecast."

Even with some art-house fare, the average haul for a best-picture nominee this year was $131 million. (*Winter's Bone, 127 Hours* and *The Kids Are All Right* are the only nominated films with comparatively low box-office draws.) But

just five years ago, when *Crash* won best picture, no movie earned more than $83 million.

"It's encouraging when a small movie like *King's Speech* does $100 million, because it relies on good story line," Gray says. "It's worth it for studios to see small movies that can do well, in terms of quality and commercially."

Analysts credit studio strategy for some of the unlikelier hits of the Oscar race.

The key to success for *Black Swan* ($101.8 million) may have been dumping the ballet shoes. Fox Searchlight pitched this Natalie Portman drama as a psychological thriller, not a dance flick. That said, "even when it was at $60 million, we didn't expect it to get this far," says Searchlight's Sheila DeLoach.

With *The King's Speech* ($104.7 million), Harvey Weinstein held onto this story of George VI's stuttering struggles until it won high-profile industry awards.

"By the time it came out, people were waiting to see it," says Jeff Bock of the industry tracking firm Exhibitor Relations. "It was classic Oscar campaigning, played perfectly."

And *True Grit* ($164.6 million) may be the most accessible film by Joel and Ethan Coen. By combining a classic title with Oscar favorites, including star Jeff Bridges, *True Grit* "became bigger than its genre. It was no longer just a Western, which was supposed to be dead," says Paramount Pictures vice chairman Rob Moore.

"That's the way all of these movies succeeded," Moore says. "They became bigger than their genres and became a movie people were talking about."

Study Questions for Case Study 3.2

"Nominees score big at box office" by Scott Bowles
USA Today
February 22, 2011

1. Compile a list of the ten films nominated for Best Picture. Include the box-office income of the seven films for which it is given.

2. Find the average (mean) and median of the box-office income of the ten films. Explain how you determined the mean and the median.

3. What was the combined box office income of the three movies for which individual incomes were not given in the article?

4. How many of these ten films earned more than the highest grossing film nominated for Best Picture five years ago? Give and explain some factors for why 2011's films earned more than 2006's films.

Case Study 3.3: Median and Mean Home Prices

Resource Material: "Real Estate Track," *The Morning News of Northwest Arkansas*, March 1, 2009, and "Median home prices fell nationwide in fourth quarter: NAR report," *The Seattle Times*, February 12, 2009.

Learning Goals: The learning goals of this case study include understanding the difference between means and medians, explaining the effect of outliers on these measures, continued work with percents and ratios, and placing large numbers in a personal context.

Articles for Case Study 3.3

Real Estate Track, *The Morning News of Northwest Arkansas*, March 1, 2009

SUNDAY, MARCH 1, 2009

REAL ESTATE TRACK

Average List Prices
Of New And Existing Homes

	Feb. 2	Feb. 9	Feb. 16	Feb. 23
Northwest Arkansas	$232,915	$232,781	$232,732	$233,934
Central Arkansas	$224,472	$224,008	$224,368	$225,877
Fort Smith/Van Buren	$185,513	$185,388	$181,138	$183,530
Jonesboro Area	$197,398	$193,290	$194,383	$192,224

Median List Prices
Of New And Existing Homes

	Feb. 2	Feb. 9	Feb. 16	Feb. 23
Northwest Arkansas	$164,200	$164,427	$164,900	$164,900
Central Arkansas	$166,900	$166,900	$167,900	$169,000
Fort Smith/Van Buren	$148,500	$149,900	$149,000	$149,900
Jonesboro Area	$149,900	$149,500	$149,900	$149,200

Inventory Of New And Existing Homes

	Feb. 2	Feb. 9	Feb. 16	Feb. 23
Northwest Arkansas	5,552	5,543	5,511	5,534
Central Arkansas	4,795	4,866	4,860	4,897
Fort Smith/Van Buren	1,088	1,091	1,093	1,092
Jonesboro Area	595	583	600	616

SOURCE: ARKANSAS REALTORS ASSOCIATION

Courtesy of Arkansas Realtors Association.

Associated Press
February 12, 2009
Median home prices fell nationwide in fourth quarter: NAR report
By ALAN ZIBEL

WASHINGTON—Home prices fell in nearly nine out of every 10 U.S. cities in the fourth quarter of last year as low-cost foreclosures flooded the market and the housing market's decline spread nationwide.

The National Association of Realtors said today that median sales prices of existing homes declined in 134 out of 153 metropolitan areas compared with the same period in 2007. Sales fell in all but six states.

Nationwide, the median sales price was $180,100, down 12 percent from a year ago. But price declines of 30 percent or more were found in much of California, plus parts of Michigan, Florida, Arizona and Nevada. The biggest drop, of more than 50 percent, was in Fort Myers, Fla.

The Northwest MLS reported earlier that the median sales price of a single family home in King County fell 7.24 percent compared to the previous December, to $403,500. The median sales price of a condo in King County came in at $288,895, down 0.38 percent compared to the previous December.

The median sales price for single family homes and condos in the nearby Snohomish, Pierce and Kitsap counties also declined year over year. In Snohomish median prices fell 9.62 percent to $307,000; Pierce dropped 12.95 percent to $235,000; and Kitsap fell 16.57 percent to $221,500.

President Barack Obama visited Fort Myers earlier this week in an effort to sell his economic rescue package, which lawmakers are preparing to send to his desk by Friday.

The states in which sales rose—Nevada, California, Arizona, Florida, Minnesota and Virginia—are places where buyers have been able to snap up foreclosed homes at a bargain. Sales more than doubled in Nevada, rose 85 percent in California, and nearly 43 percent in Arizona.

"We see a pattern of strong sales gains, particularly in lower price homes, in areas with price declines resulting from foreclosures," Lawrence Yun, the trade group's chief economist, said in a prepared statement.

In California and Florida, sales of distressed properties accounted for about two-thirds of all sales, compared with about 45 percent nationally.

A nasty brew of strict lending standards, falling home values, soaring foreclosures and a severe recession is filtering through the housing market.

Nationwide, more than 274,000 homes received at least one foreclosure-related notice in January, according to RealtyTrac Inc., an Irvine, Calif.-based foreclosure listing service. That was down 10 percent from December, but still up 18 percent from the same month a year ago. The numbers would have been higher if not for efforts to stall the foreclosure process.

More than 2 million American homeowners faced foreclosure proceedings last year, and that number could soar as high as 10 million in the coming years depending on the severity of the recession, according to a report last month by Credit Suisse.

Study Questions for Case Study 3.3

"Real Estate Track"
The Morning News of Northwest Arkansas
March 1, 2009

1. The information in "Real Estate Track" uses both means and medians to describe home prices for different regions of Arkansas at different times in February.

    a. What is the total of all prices for listed homes for Northwest Arkansas as of February 23?

    b. How many homes in Northwest Arkansas were listed for less than $164,900 on February 23?

    c. The values in the second table (Median List Prices) appear to be lower than those in the first table (Average List Prices). Explain why this is to be expected.

    d. Which region of Arkansas saw the greatest change in home prices from February 2 to February 23? Describe carefully how you decided to measure the change in home prices. Be sure to support your conclusion by carefully describing how you measured "change" in each region and why this "change" is the greatest.

    e. Which region of Arkansas saw the least change in home prices from February 2 to February 23? Be sure to support your conclusion by carefully describing how you measured "change" in each region and why this "change" is the least.

    f. Describe how mean and median home prices changed during February for the Fort Smith/Van Buren region. Explain any unusual findings.

    g. On a hypothetical set of home prices, investigate how the mean and median behave when a few home prices change dramatically.

    h. Which measure of center, mean or median, would typically be used to describe home prices? Why?

"Median home prices fell nationwide in fourth quarter: NAR report" by Alan Zibel
The Seattle Times
February 12, 2009

2. Alan Zibel's article "Median home prices fell nationwide in fourth quarter" gives some additional information regarding home prices in the U.S.

   a. Each of the first two sentences of the article gives some information about falling home prices in the U.S. Is the information in these sentences consistent? Explain your reasoning.

   b. According to this report, what was the median sales price for homes in the U.S. "a year ago"?

   c. How many homes received at least one foreclosure-related notice in the U.S. in January 2009? How many the previous December? The previous year?

   d. Give an argument to explain why sales increased more in Nevada than California.

   e. Give another argument to explain why sales increased more in California than in Nevada.

   f. The states mentioned in the article associated with large drops in home prices are also the states mentioned as those in which "sales rose."

      i. Which states are these?

      ii. One might argue that these two facts are inconsistent. Indeed, home prices should drop when there is not much activity on the housing market. (Houses are not selling, so sellers lower their prices to encourage the few buyers out there to buy their house.) Based on the information in the article, why are these two facts not inconsistent?

   g. If your community was similar to the national picture, how many homes would have faced foreclosure proceedings last year? If the estimate in the article is correct, how would this number change "in the coming years"?

Introduction to Indices (also Indexes)

Definition. An **index number** (or simply an **index**) for a measurement is the product of the ratio of the measurement's value to a reference (fixed) value for the measurement and the base value of the index. That is:

$$\text{Value of index} = \frac{\text{Value}}{\text{Reference Value}} \times \text{base}.$$

The base is the value of the index for the reference value. Often 100 is chosen as the base so that the value of the index at the reference value is 100. If one wants that to be some other number, say 20, then 20 would replace 100. When 100 is the base value, the index is the percent the value is of the reference value, often without saying percent or using the % sign.

The reference value may be the value of the measurement at a particular time or place or it may be an average (arithmetic mean) of the values of the measurement. Indexes (or indices) are ways of comparing values (say, over time or place) by using a fixed reference value.

As an example, consider the price of gasoline (per US gallon) as given in the table below with the price in 1975 being chosen as the reference value. Note that there are two columns for the Price Index, one with a base value of 100 and one with a base value of 20. Again, in practice, any base value can be chosen.

Year	Price	% of 1975 price	Price Index (1975 = 20)	Price Index (1975 = 100)
1955	29.1¢	51.3%		
1965	31.2¢			
1975	56.7¢	100%	20	100
1985	119.0¢			
1995	120.0¢			
2000	155.0¢			
2004	190.0¢			
2007	280.0¢			
2011	340.0¢			

Exercise: Fill in the values in the above table.

Caution: Some quantities or measures are called indexes but do not satisfy the above definition. Most such quantities are simply measures or condensed measures.

Definition. A measure that partially describes a set or the items in a set is sometimes called a **condensed measure**.

For example, the mean and median of a set of numbers are condensed measures. The weight or height of a person is a condensed measure.

Example 1. Body Mass Index (BMI). Some states now require reporting of BMI for public school students because the BMI is a measure that has been related to good health. The BMI is computed as follows:

$$BMI = 703\frac{w}{h^2}$$ where w is in pounds and h is in inches.

Using the definition above of an index, is the BMI an index? Why or why not?

Example 2. Consumer Price Index (CPI). The Consumer Price Index (CPI) is a measure of the average change over time in the prices paid by urban consumers for a market basket of consumer goods and services. The US Department of Labor provides several types of consumer price indices, by geographic area and by types of items included. The CPI considers items in the following categories:

- FOOD AND BEVERAGES (breakfast cereal, milk, coffee, chicken, wine, service meals and snacks)
- HOUSING (rent of primary residence, owners' equivalent rent, fuel oil, bedroom furniture)
- APPAREL (men's shirts and sweaters, women's dresses, jewelry)
- TRANSPORTATION (new vehicles, airline fares, gasoline, motor vehicle insurance)
- MEDICAL CARE (prescription drugs and medical supplies, physicians' services, eyeglasses and eye care, hospital services)
- RECREATION (televisions, pets and pet products, sports equipment, admissions)
- EDUCATION AND COMMUNICATION (college tuition, postage, telephone services, computer software and accessories)
- OTHER GOODS AND SERVICES (tobacco and smoking products, haircuts and other personal services, funeral expenses)

The CPI differs from a cost of living index (Example 3 below) in that the CPI does not cover all items that contribute to cost of living such as taxes, climate, and safety issues.

The current CPI has a base of 100 referenced to the 1982–84 period. Thus a CPI of 181 in 2004 in the urban south means that the cost of goods purchased in this area in 2004 cost approximately 181% what the same goods cost in 1982–84.

Example 3. Cost of Living Index (COLI). One cost of living index takes an average cost of living across the US as its base of 100 or 100%. The time period is fixed, that is, the COLI compares different locations at a fixed time. The index varies by geographical location, e.g., cities, giving the percent of the national average for each location. A composite index of 92.2 for Memphis, TN, means that the cost of living in Memphis is 92.2% of the national average. An index of 148.2 for Seattle, WA, means that the cost of living in Seattle is 148.2% of the national average.

Example 4. Standard & Poor's 500 Stocks Composite Average (S&P 500). This is an index although it is often called an average. The S&P 500 is an index based on the market capitalization (number of shares times the value of a share) of 500 stocks. The base period for the S&P 500 is 1941–43, which was assigned a value of 10. Consequently

$$S\ \&\ P\ 500 = \frac{\text{sum of capitalizations in current period}}{\text{sum of capitalizations in 1941–43}} \times 10.$$

Example 5. Dow Jones Industrial Average (DJIA). This widely quoted measure is not an index, but an unusual average of the prices of 30 stocks (since 1928), all very large companies. In 2007, the DJIA was over 14,000, but was back near 7,000 in 2009, either of which is quite a large number that is purported to be the average of the price of 30 stocks. We would expect such an average to be less than $100 since most shares of stock have prices under $100. Here is an example that illustrates how the DJIA is computed.

Definition: A *stock split* is a process by which a stock price is lowered by issuing more shares at a lower price. For example, if you own 50 shares of a stock that is selling for $90 per share and there is a 2-for-1 stock split, your 50 shares @ $90 are replaced by 100 shares @ $45. If it were a 3-for-1 split, your shares would be replaced by 150 shares @ $30.

We begin our own stock average with three stocks: A, B, & C. At first, the stocks sell for $20, $30, and $10. The first value of our average is $\frac{\$20 + \$30 + \$10}{3} = \$20$. One year later we still have the same stocks and there have been no stock splits (see definition above), but the value of stock has changed. Assume the prices are now $25, $40, and $12. So our average is $\frac{\$25 + \$40 + \$12}{3} = \$25.67$. The $40 stock splits 2 for 1 and the new price is $20. Now we are averaging $25, $20, and $12 but the average should be 25.67, so we compute a divisor (other than 3) that will make the average $25.67.

$$\frac{\$25 + \$20 + \$12}{d} = 25.67$$

Therefore, one must use a divisor of $d = 2.22$.

Henceforth we use this divisor until some other artificial change in the price of one of our stocks (such as a split) causes us to re-compute the divisor. Another example of an artificial change would be the replacement of one of our stocks with a different stock with a different value. The DJIA divisor has changed many times since its original value of 30. In 2011 it was approximately 0.13.

Warm Up Exercises on Indices

1. The cost of a loaf of bread was $0.45 in 1975. Use 1975 as a base year and fill in the four blank cells in the following cost-of-bread index (CBI) table.

Year	Cost of loaf of bread	CBI
1965	$0.35	
1975	$0.45	100
1985	$0.60	
1995		175
2011	$1.50	

2. If the average US salary is $41000 per year, what are the comparable salaries in the three cities in the table when considering cost of living?

City	Cost of Living Index	Salary
Farmville	84	
Metropolis	125	
Midville	98	

Case Study 3.4: Body Mass Index

Resource Material: "Tell the Truth: Does this Index Make Me Look Fat" by Gina Kolata, *New York Times*, November 28, 2004.

Learning Goals: The learning goals of this case study include understanding the Body Mass Index (BMI), distinguishing a condensed measure from an index, and computing BMIs in both metric and customary units.

The table accompanying this article gives the formula for the Body Mass Index (BMI) and gives the BMI for 13 US Presidents, listed in order of increasing BMI. The article discusses the meaning and use of the BMI and how the cut points for overweight and obese are determined.

Warm Up Exercises for Case Study 3.4

1. Body Mass Index (BMI) of a person is computed as follows:

    $$BMI = 703 \frac{\text{weight in pounds}}{(\text{height in inches})^2}.$$

    a. Compute the BMI of a person who is 5 feet 5 inches tall and weighs 125 pounds.

    b. Convert the 5 feet 5 inches into meters (1 meter = 39.37 inches) and 125 pounds into kilograms (1 kilogram = 2.205 pounds) and compute the BMI again using

    $$BMI = \frac{\text{weight in kilograms}}{(\text{height in meters})^2}$$

2. Answer the following:

    a. Compute the BMI of a person who is 1.6 meters tall and weighs 80 kilograms.

    b. How tall (in inches) is a person who has a BMI of 25 and weighs 160 pounds?

Article for Case Study 3.4

New York Times
November 28, 2004
Tell the Truth: Does This Index Make Me Look Fat?
By GINA KOLATA

THERE'S an overweight man in the White House and his name is George W. Bush.

Yes, the president of the United States, known for his robust good health, is officially overweight, according to the standards of the National Institutes of Health. At 6 feet and 194 pounds, his body mass index, or B.M.I., a measurement of height relative to weight, is 26.4, and 25 or above is officially overweight for both sexes.

Oval Office?

Almost half of American presidents have been overweight, based on the body mass index guidelines established by the Center for Disease Control and Prevention. At the extreme ends of the spectrum: James Madison and William Howard Taft. Here is a selection:

Being over weight is defined as a body mass index of 25 or higher. Body mass index is determined by using this formula:

$$B.M.I = \left(\frac{\text{WEIGHT IN POUNDS}}{(\text{HEIGHT IN INCHES})^2} \right) \times 703$$

Source: National Health and Nutrition Survey

		HEIGHT	WEIGHT	B.M.I.
James Madison	1809–17	5 feet 4 inches	99	17.0
Andrew Jackson	1829–37	6 feet 1 inch	144	19.0
Abraham Lincoln	1861–65	6 feet 4 inches	178	21.7
John F. Kennedy	1961–63	6 feet	167	22.6
Ulysses S. Grant	1869–77	5 feet 8.5 inches	156	23.3
Richard M. Nixon	1969–74	5 feet 11.5 inches	174	23.9
Millard Fillmore	1850–53	5 feet 9 inches	164	24.2
George Washington	1789–97	6 feet 2 inches	199	25.5
George W. Bush	2001–	6 feet	194	26.3
Bill Clinton	1993–2001	6 feet 2.5 inches	223	28.3
Chester A. Arthur	1881–85	6 feet 2 inches	224	28.7
Theodore Roosevelt	1901–09	5 feet 10 inches	210	30.2
William Howard Taft	1909–13	6 feet	312	42.3

And so President Bush joins about 65 percent of Americans who are overweight or obese—a status derived solely from that body mass index dividing line of 25.

Of course, the authorities can be wrong when it comes to matters of weight and health. Just last week, for example, the Centers for Disease Control and Prevention said its previous estimate that 400,000 Americans die each year from obesity and overweight was too high.

So, does President Bush's B.M.I. of 26.4 actually make much difference to his health? "Probably not," says Dr. George Bray, an obesity researcher at the Pennington Biomedical Research Center of Louisiana State University. "Body mass is an index from which you start to make an evaluation of an individual."

"The meaning of B.M.I. has to be modulated by other factors, including age, gender, physical activity, race and central fat distribution," he added, referring to the amount of fat a person carries around the waist and abdomen.

The National Heart, Lung and Blood Institute agrees, and includes high blood pressure and cigarette smoking as factors in determining whether weight is a problem.

Dr. Eric Oliver, a political scientist at the University of Chicago who is writing a book about the politics behind the obesity epidemic, says that the more one looks into the health claims behind the overweight designation, the more arbitrary they seem. "From a scientific perspective, there is no way you could make those claims based on the data," he said.

Before 1998, a man was officially overweight with a body mass index of 27.8 and a woman at 27.3—numbers based on the body mass averages of people in their 20's. But a committee convened by the national heart institute redefined overweight to be a body mass index of 25 or over, for men and women.

As justification, it cited studies finding a slight increase in death rates as the body mass increased above 25. The increases tended to be modest up to an index level of 30, the official designation for obesity. And though this mortality data was hardly solid, Dr. Oliver said, "Overnight, 37 million people were suddenly overweight."

Dr. Katherine Flegal, a statistician at the National Center for Health Statistics, says there is a problem with the graph, which was shaped like a shallow U with 25 at its base. "If the nadir is 25, then, yes, mortality does begin to increase with B.M.I.'s above 25, so the statement is literally true," she said. "But of course if the curve is U-shaped with a nadir at 25, then mortality also increases as B.M.I. decreases below 25." All of which suggests that a body mass index of 25, far from being dangerous, is actually optimal.

So perhaps it's a good thing that few people seem to assess their weight based on the B.M.I.

On one hand, "almost no women think they are skinny," Dr. Flegal. On the other, she said, only 42 percent of men with a B.M.I. of 25 think they are overweight.

Study Questions for Case Study 3.4

"Tell the Truth: Does This Index Make Me Look Fat?" by Gina Kolata
New York Times
November 28, 2004

1. Does the BMI satisfy the conditions of the definition of an index? Does it allow comparisons of a BMI to some base BMI? Support your answers.

2. Check the BMI values for James Madison and William Howard Taft given in the article.

3. Use 2.205 pounds = 1 kilogram and 39.37 inches = 1 meter and express Bill Clinton's height in meters and weight in kilograms.

4. Compute Bill Clinton's BMI in pounds/inches squared and compare this to the calculation $\dfrac{\text{weight in kg}}{(\text{height in m})^2}$.

5. Based on the results of #5 above, where does the 703 arise in computing the BMI?

6. Before 1998, what range of BMI values was accepted as indication of being overweight?

7. Why was the cutoff BMI for being overweight reduced to 25?

8. Why was the cutoff for obesity established at 30?

9. Draw a portion of a graph of BMI versus mortality rates that satisfies the description in the article.

10. What does nadir mean?

11. Explain, in terms of your graph from #9, what Dr. Katherine Flegal is observing about the data on mortality versus BMI.

12. How does Barack Obama's BMI compare to those in the table? Make a guess and then find his weight and height and compute his BMI.

Case Study 3.5: Stock Market

Resource Material: "Software Sector Snapshot," *New York Times*, September 22, 2007.

Learning Goals: The learning goals of this case study include understanding how some stock averages and indices are computed.

The stock market in the US is extensive and complex, and there are many indices and averages that measure the changes in the prices of stocks in many different segments of the market. The two major US stock exchanges are the New York Stock Exchange (NYSE) and the National Association of Securities Dealers Automated Quotation system (NASDAQ). At the beginning of 2011, there were more than 3,000 companies on the NYSE with a capitalization (the product of the number of shares of stock with the price of a share) of approximately $13 trillion, and approximately 2,700 companies on NASDAQ with an approximate capitalization of $3.5 trillion. So, on the two large US markets, there were about 6,000 companies worth approximately $16.5 trillion. Three major measures of US corporate economic health are the Dow Jones Industrial Average (DJIA), the Standard and Poor's 500 Index (S&P500), and the NASDAQ Composite Index (NASDAQ). There are 30 companies on the DJIA with a combined value of approximately $2.5 trillion, 500 companies on the S&P500 with a combined value of about $10 trillion, approximately 2,700 on NASDAQ with a combined value of $3.5 trillion.

Prices of stocks are tracked in groups of stocks such as energy stocks, retailing, telecom stocks, industrials, and software and services. The graphic under study here indicates how the software and services sector of the market performed over the previous week and over the previous year as of the date of the graph, September 22, 2007.

Warm Up Exercises for Case Study 3.5

1. The Four Stock Index (FSI) is computed as the S&P 500 is computed. The base year is 1980 and the four companies had the situations below in 1980 and 2012. If the initial FSI was set at 15, fill in the capitalizations and compute the FSI in 2012.

Company	1980			2012		
	Shares	Price	Capitalization	Shares	Price	Capitalization
A	5000	$12		10000	$20	
B	2500	$20		3000	$28	
C	1000	$80		3500	$23	
D	9000	$30		17000	$25	

2. The Four Stock Average (FSA) begins with four stocks, A, B, C, and D, and will compute the FSA the way the DJIA is computed. At the beginning the stocks sell for $16, $20, $41, and $45 per share.

   a. Compute the initial FSA.

   b. One year later there have been no stock splits and the stocks sell for $18, $29, $51, and $54 when the $54 stocks split 3 for 1 and the new price per share is $18. Find the FSA and the new FSA divisor.

   c. Another year later there have been no additional splits and the stocks sell for $20, $33, $49, and $26. Find the FSA.

Article for Case Study 3.5

THE NEW YORK TIMES, SATURDAY, SEPTEMBER 22, 2007

MARKET GAUGES

S.&P. 500 ↗ 1,525.75 +7.00	**DOW INDUSTRIALS** ↗ 13,820.19 +53.49	**NASDAQ COMPOSITE** ↗ 2,571.22 +16.93	**10-YEAR TREASURY YIELD** ↘ 4.62% −0.08	**CRUDE OIL** ↘ $81.62 −$0.16	

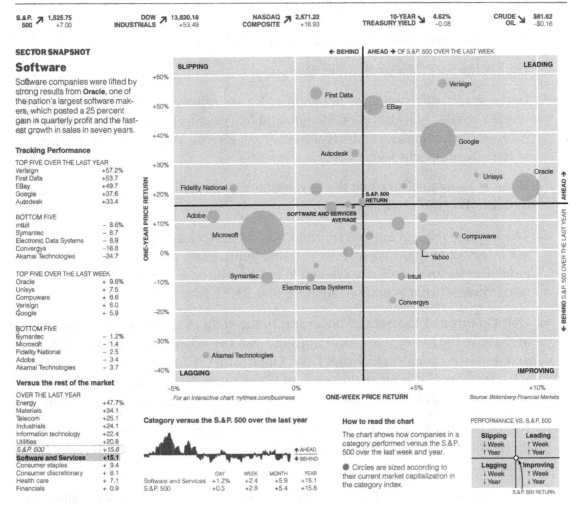

SECTOR SNAPSHOT

Software

Software companies were lifted by strong results from **Oracle**, one of the nation's largest software makers, which posted a 25 percent gain in quarterly profit and the fastest growth in sales in seven years.

Tracking Performance

TOP FIVE OVER THE LAST YEAR

Verisign	+57.2%
First Data	+53.7
EBay	+49.7
Google	+37.6
Autodesk	+33.4

BOTTOM FIVE

Intuit	− 8.6%
Symantec	− 8.7
Electronic Data Systems	− 8.9
Convergys	−16.8
Akamai Technologies	−34.7

TOP FIVE OVER THE LAST WEEK

Oracle	+ 9.6%
Unisys	+ 7.5
Compuware	+ 6.6
Verisign	+ 6.0
Google	+ 5.9

BOTTOM FIVE

Symantec	− 1.2%
Microsoft	− 1.4
Fidelity National	− 2.5
Adobe	− 3.4
Akamai Technologies	− 3.7

Versus the rest of the market

OVER THE LAST YEAR

Energy	+47.7%
Materials	+34.1
Telecom	+25.1
Industrials	+24.1
Information technology	+22.4
Utilities	+20.8
S.&P. 500	*+15.8*
Software and Services	**+15.1**
Consumer staples	+ 9.4
Consumer discretionary	+ 8.1
Health care	+ 7.1
Financials	+ 0.9

For an interactive chart: nytimes.com/business

ONE-WEEK PRICE RETURN

Source: Bloomberg Financial Markets

Category versus the S.&P. 500 over the last year

↑ AHEAD
↓ BEHIND

	DAY	WEEK	MONTH	YEAR
Software and Services	+1.2%	+2.4	+5.9	+15.1
S.&P. 500	+0.5	+2.8	+5.4	+15.8

How to read the chart

The chart shows how companies in a category performed versus the S.&P. 500 over the last week and year.

● Circles are sized according to their current market capitalization in the category index.

PERFORMANCE VS. S.&P. 500

Slipping ↓ Week ↑ Year	**Leading** ↑ Week ↑ Year
Lagging ↓ Week ↓ Year	**Improving** ↑ Week ↓ Year

S.&P. 500 RETURN

Study Questions for Case Study 3.5

"Software Sector Snapshot"
New York Times
September 22, 2007

1. Why is the Dow Jones Industrial Average (DJIA) not called an index?

2. How is the S&P 500 stock index computed?

3. In the graph, different software companies are represented by different sized shaded circles. For examples, Google and Microsoft have large circles and Akamai Technologies and Unisys have small circles. What is the significance of the size of the circles?

4. What is being measured on the horizontal axis of the graph? What is being measured on the vertical axis?

5. The heavier vertical and horizontal lines, which usually intersect at (0,0), do not intersect at (0,0). We usually call this point, (0,0), the origin of the rectangular Cartesian coordinate system. What are the "coordinates" of this translated "origin" that is marked by S&P 500 RETURN?

6. Call Quadrant 1 the area of the graph to the right of the heavy vertical line and above the heavy horizontal line, and call the other three quadrants 2, 3, and 4 moving counterclockwise from Quadrant 1. Explain how the company is performing if it is located in each of the quadrants. Namely, what does it mean if a company is in Quadrant 1? Quadrant 2? Quadrant 3? Quadrant 4?

7. Notice that the scale on the horizontal axis is different from the scale on the vertical axis. Compare the two scales and explain why this difference is needed.

8. How has the category of software and services fared versus the S&P 500 over the past year?

9. If you could go back in time to September 15, 2007, and invest in this market sector, in which company would you invest? What company would have been the worst choice? If you had invested $1000 on September 15, 2007, in each of these two companies, how much would you have gained or lost on each by September 22?

10. If you could go back in time to September 22, 2006, and invest in this market sector, in which company would you invest? What company would have been the worst choice? If you had invested $1000 on September 22, 2006, in each of these two companies, how much would you have gained or lost on each by September 22, 2007?

Case Study 3.6: Chuck Taylor All Star

> Resource Materials: Advertisement from the internet for Chuck Taylor All Star canvas shoes.

Learning Goals: The learning goals of this case study include finding ways to measure the value of the dollar over time using the cost of a familiar object whose value has been reasonably constant over time.

Warm Up Exercises for Case Study 3.6

1. The price of a loaf of bread was $0.70 in 1975 and $1.75 in 2011. Based on this information find the following:
   a. The value of the 2011 dollar in 1975 dollars.
   b. The value of the 1975 dollar in 2011 dollars.
2. Using the results of #1 above, find the approximate cost of a gallon of gasoline in 1975 if a gallon of gasoline costs $3.50 in 2011.

Article for Case Study 3.6

Study Questions for Case Study 3.6

1. The consumer price index (CPI) was approximately 220 at the beginning of 2011 when the advertisement above appeared. The CPI at the beginning of 1950 was approximately 24. Based on this information, find the price of the Chuck Taylor All Star shoe in 1950.

2. The Chuck Taylor All Star shoe was introduced under the name Chuck Taylor in 1923. What was the CPI of that year? Based on this information find the introductory price of the shoe.

3. If the Chuck Taylor All Star shoe sold for $17 in 1980, find the value of the 2011 dollar in 1980 dollars.

4. What characteristics of the Chuck Taylor All Star shoe make it a reasonable product on which to base the value of the dollar?

5. What are two other products whose prices would be reasonable indicators of the value of the dollar over the past 50 years? Why?

6. What are two products whose prices would not be reasonable indicators of the value of the dollar over the past 50 years? Why?

Case Study 3.7: Measuring Spending and Revenue in Different Units

Resource Material: "GOP Disputes," Associated Press Graphic, March 4, 2003, and "Is the glass half full . . . ," *Arkansas Democrat-Gazette* Graphic, May 18, 2003.

Learning Goals: The learning goals of this case study include analyzing quantitative arguments, measuring government revenue in different units, and gleaning additional information from multiple graphs.

Units for Measuring Government Revenue and Spending

In this case study, the different units are more complex than feet and inches or pounds and kilograms. In this case study we examine different ways of measuring and comparing government spending and revenues. If one wants to compare government spending over time, one must be able to compare dollar amounts from different time periods. Since the buying power of the dollar changes over time, the value of $1 also changes. Therefore the unit of "dollar" is ambiguous. We need to introduce different units to help us compare government spending or revenue amounts from different time periods. To complicate matters further, in any one year one may not wish to use "dollars" at all to measure a quantity such as government spending. Rather, one may wish to measure government spending by comparing it to another quantity. This is analogous to measuring an individual's annual spending not in dollars but in terms of their annual salary. For example, rather than saying that Gillian spends $33,000 on rent, utilities, groceries, and gas in a single year, one might report this amount as 82% of her annual income. The three different units used in this case study are nominal dollars, constant dollars, and percentage of gross domestic product (GDP).

Nominal dollars[4] is an expression used to measure an amount of money in current dollars, or more suggestive, dollars-of-the-day or dollars-of-the-year. When expressed in nominal dollars, quantities can be very difficult to compare! For example, in 1957 the cost of a gallon of gasoline was approximately $0.25. Fifty years later, in 2007, the cost of a gallon of gasoline was approximately $3.00. These two amounts are in nominal dollars. Did gasoline cost 12 times as much in 2007 as in 1957? To understand how these two costs for the same product at different times compare one has to know how the nominal 1957 dollar compares to the nominal 2007 dollar.

Constant dollars differ from nominal dollars because of inflation and deflation, mostly inflation in recent years. Inflation of 5% over the year 2007 means that in general everything costs 5% more at the end of 2007 than at the beginning; if a loaf of bread costs $2 on January 1, 2007, a comparable loaf of bread costs $2.10 on January 1, 2008. (The dollar's value changes during the year, but we choose the value at some instant in the year and call that the dollar's value for the year.) Consequently the dollar in 2008 (commonly called the 2008 dollar) is worth less than the 2007 dollar. (We will use the notation

---

4. It is noted that the first graphic in this case study calls nominal dollars or dollars-of-the-day by real dollars. In fact, real dollars is sometimes used as a synonym for constant dollars.

(2008-dollar) to avoid confusing numbers with units.) There are two ways of comparing (2007-dollars) and (2008-dollars), assuming an inflation of 5% over 2007:

One (2007-dollar) is equivalent to 1.05 (2008-dollars).

One (2008-dollar) is equivalent to 0.95238 (2007-dollars). *Be careful! It would be incorrect to say that one (2008-dollar) is equivalent to 0.95 (2007-dollars).This is because 0.95238 increased by 5% is 1, while 0.95 increased by 5% is 0.9975.*

As a consequence of the changing value of the dollar over time, to compare spending or revenues over time, one often fixes a year and expresses the spending or revenue in the dollar of the fixed year, that is, in terms of a constant dollar.

The third way to measure spending and revenues is as a percentage of the Gross Domestic Product (GDP), which is the total value of goods and services produced in the country in a given year. The rationale for using percentage of GDP as a unit can be illustrated as follows: If the country spent 5% of the GDP for defense in 1950 then the same level of spending in 2000 would be 5% of the GDP in 2000. A personal analogy would be spending a fixed percentage of one's income for transportation.

Issues Regarding the Measurement of Government Revenue and Spending
The two graphics under study here, appearing about 2.5 months apart in 2003, contain five graphs. Three of the graphs under the headline *GOP Disputes* show the federal budgets and surpluses or deficits for years 1980 to 2003, first in nominal dollars (i.e., the dollars of the years), then in constant (1996-dollars), and finally as a percentage of GDP (of the various years). These graphs are the basis of the argument as to whether or not the projected deficit in 2003 would be a record deficit. There is one incidental instance of measuring in different units in this graphic. It involves measuring in billions of dollars and in trillions of dollars.

The first of the other two graphs under the headline *Is the glass half full* . . . shows the federal government's discretionary spending (as opposed to mandatory spending for programs such as social security and medicare) in nominal dollars for the years 1996–2003. The second graph here gives government revenues as a share of GDP for the years 1996–2003. The two *Is the glass half full* . . . graphs are the basis for the argument as to whether or not we should be spending more or less on discretionary programs such as education, defense, and transportation. As we shall see, the issue of which units to use plays an important role in analyzing these budget situations.

In discussing the meanings of these graphs and their importance to the arguments involved, we can assume that government spending is either discretionary or mandatory; that is, the sum of discretionary and mandatory spending is total spending. One other slightly subtle point involves Study Question #2a below. This question might seem incorrectly stated as decreasing government revenue should support decreased spending. However, there is another view that is most likely the reason liberals point to decreasing government revenues.

The study questions involve analyses of the five graphs, sometimes separately and sometimes together. The two graphics with the five graphs are on the following pages.

Warm Up Exercises for Case Study 3.7

1. If you spent $100 per month for transportation when your salary was $26,000 per year, compute the percentage of your salary used for transportation. If your salary increased to $35,000 and you allocated the same percent of salary for transportation, how much per month would you allocate to transportation?

2. Assume that a loaf of bread cost $0.25 in 1950 and an equivalent loaf of bread cost $2.25 in 2000. Based on the cost of this loaf of bread, what is the value of a 1950-dollar in 2000-dollars?

Articles for Case Study 3.7

GOP Disputes

Vying To Set Deficit Record Straight

The $304 trillion budget deficit projected by the Bush administration
for 2003 would be the largest ever. But Republicans are quick to
point out the deficit would not be a record when adjusted for inflation
or represented as a percentage of the gross domestic product.

Federal Budget Surpluses And Deficits

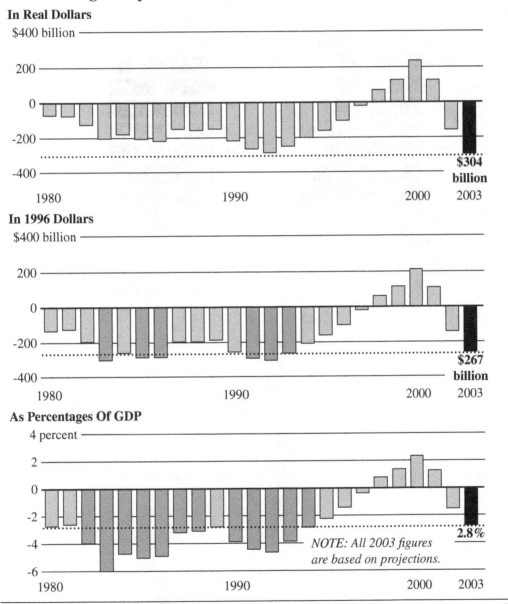

In Real Dollars

In 1996 Dollars

As Percentages Of GDP

*NOTE: All 2003 figures
are based on projections.*

SOURCE: Office of Management and Budget   AP

© AP/WORLD WIDE PHOTOS Used with permission

Courtesy of the Associated Press.

Arkansas Democrat ⁂ Gazette   18 May 2003

Is the glass half full . . .

Conservatives look at the federal government and see an alarming growth in discretionary spending—spending related to mandatory programs such as Social Security and Medicare:	*Liberals look at the federal government and see an alarming decline in government revenues as a share of gross domestic product—the total value of the goods and services produced in the country.*

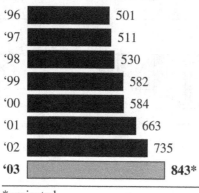

Discretionary spending
IN BILLIONS OF DOLLARS

'96	501
'97	511
'98	530
'99	582
'00	584
'01	663
'02	735
'03	843*

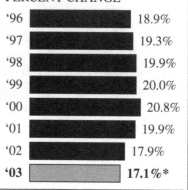

Government revenues as a share of GDP
PERCENT CHANGE

'96	18.9%
'97	19.3%
'98	19.9%
'99	20.0%
'00	20.8%
'01	19.9%
'02	17.9%
'03	17.1%*

*projected

SOURCES: Heritage Foundation. Center on Budget and Policy Priorities

Arkansas Democrat-Gazette

Study Questions for Case Study 3.7

"GOP Disputes"
Morning News of Northwest Arkansas
March 4, 2003

At some point in 2003, the $304 billion federal budget deficit projected by the Bush administration would be the largest ever. However, Republicans argued that it was not a record deficit if one measured this projected deficit and the deficits since 1980 in either constant 1996-dollars or as percentages of the Gross Domestic Product (GDP).

1. **Answer the following:**

    a. **At one point in this graphic (headlined by GOP Disputes), the projected deficit is incorrectly stated as $304 trillion. Locate this error and express $304 billion measured in trillions.**

    b. **In what years was there a budget surplus? In what year was the largest budget surplus?**

    c. **From the information in the three graphs, what is the 2003 dollar's value in (1996-dollars)? Explain your method.**

    d. **From the information in the three graphs, what is the 2003 GDP measured in (2003-dollars)? Explain your method.**

    e. **From the information in the three graphs, what is the 1983 GDP measured in (1983-dollars)? Explain your method.**

"Is the glass half full . . ."
Arkansas Democrat-Gazette
May 18, 2003

In the situation of #1 (above) Republicans wanted to measure the projected deficit in either (1996-dollars) or a percentage of GDP. In *Is the glass half full* . . . the Republicans (conservatives) and Democrats (liberals) switch sides on this argument. In this case, measuring in (2003-dollars), the conservatives see an alarming growth in discretionary spending by the federal government. On the other hand, the liberals see a decline in government revenue when measured as a percent of GDP. Consequently, the liberals, who support increased discretionary spending, argue that the government revenues are actually declining. They would support no decline or an increase in revenues so that money would be available for discretionary spending.

2. **Answer the following:**

   a. **Explain why one cannot determine either federal government spending or federal government revenue from these two graphs alone.**

   b. **Using the information from the three graphs in #1 (above) plus the information in the two graphs in this item (#2), what is the projected federal government revenue for 2003? Explain your method.**

   c. **Find the mandatory government spending for 2003. Explain your method.**

   d. **Find the amount of 2003 discretionary spending in (1996-dollars). Explain your method.**

Case Study 3.8: FOXTROT Cartoon

Resource Material: "FOXTROT Cartoon" by Bill Amend, Universal Press Syndicate, 2004.

Learning Goal: The learning goal of this case study is unit conversion by using unit analysis.

Warm Up Exercises for Case Study 3.8
Introduction to Unit Conversion and Exercises

In order to compare quantities, they must have the same units. It is therefore important to be able to convert various quantities to different units.

To do so, we first must know a conversion factor that offers a statement of equality between the two units. For example, we know that 1 foot = 12 inches. This is an example of a conversion factor.

1. Give the conversion factor that provides an equal quantity across the various units.

   a. 1 yard   = _____ feet

   b. 1 mile   = _____ feet

   c. 1 pound   = _____ ounces

   d. 1 ton   = _____ pounds

   e. 1 kilometer   = _____ meters

   f. 1 meter   = _____ millimeters

   g. 1 meter   = _____ centimeters

   h. 1 centimeter   = _____ millimeters

   i. 1 gram   = _____ milligrams

   j. 1 square foot   = _____ square inches

   k. 1 cubic foot   = _____ cubic inches

Note that the conversion factors in *a–d* above involve units used in the English/standard system (also called the customary system). This system uses units such as miles, feet, gallons, and pounds. The conversion factors in *e–i* involve units used in the metric system. This system uses units such as liters, grams and meters. Note the importance of the prefix in the metric system.

2. What do the following prefixes mean when used in the metric system?

   Kilo: Centi: Milli:

3. The conversion factor in part *j* involves units used to measure area, while the conversion factor in part *k* involves units used to measure volume.

   a. How is the conversion factor for area involving square feet and square inches related to the previous conversion factor relating feet and inches (lengths)?

   b. How is the conversion factor for cubic feet and cubic inches (units for volume) related to the conversion factor relating feet and inches?

Once you are familiar with conversion factors, you then can use them to convert between units.

Example 1: 168 inches = ? feet

Solution: We know that 12 inches = 1 foot. To convert inches to feet, we can use what is called unit analysis (or dimensional analysis) to convert these units of measure. Unit analysis uses ratios to convert units:

$$168 \text{ in} \times \frac{1 \text{ ft}}{12 \text{ in}} = \frac{168}{12} \text{ ft} = 14 \text{ ft}. \text{ So } 168 \text{ inches} = 14 \text{ feet}.$$

Note that when we multiplied by $\frac{1 \text{ ft}}{12 \text{ in}}$, we were multiplying by a fraction equivalent to 1 (e.g., the numerator is equivalent to the denominator). You may need more than one conversion factor in order to convert units (e.g., to convert 7 days to seconds, you may need the conversion factors 1 day = 24 hours; 1 hour = 60 minutes; and 1 minute = 60 seconds to complete the conversion).

Also note that the placement of 1 foot and 12 inches in the fraction above is meaningful. We began with inches as our unit. In order to cancel inches, we needed to place inches in the denominator of the fraction, thereby placing feet in the numerator. Had we been converting 14 feet to inches, we would have used the same conversion factor, only placing 1 foot in the denominator and 12 inches in the numerator.

4. Convert the following units of measure.

    a. 45 miles   = _____ feet

    b. 345 feet   = _____ yards

    c. 7542 meters   = _____ kilometers

    d. 5 kilometers   = _____ centimeters

    e. 7 miles   = _____ inches

    f. 1296 square inches = _____ square feet

    g. 1 year   = _____ seconds

    h. 70 miles/hour   = _____ feet/minute

    i. 15 miles/hour   = _____ feet/second

    j. $1\dfrac{\text{pound}}{\text{square inch}}$ = $\dfrac{\text{kilogram}}{\text{square meter}}$

Article for Case Study 3.8

Courtesy of Bill Amend/Universal UClick.

FoxTrot Cartoon
Bill Amend

Study Questions for Case Study 3.8

1. State the problem that Jason has posed in this comic strip.

2. What conversion fact does Jason state for converting the half teaspoon of baking soda? Find out if the conversion is correct.

3. Sketch and label the box of soda that Jason describes.

4. Find the volume of soda in the box in cubic inches.

5. Density is weight per unit volume. Find the density of the soda in pounds per cubic inch and compare your results to Jason's.

6. Find the conversion factors to change cubic inches to cubic centimeters and pounds to kilograms. How are the conversion factors from inches to centimeters and from cubic inches to cubic centimeters related? Convert the density from pounds per cubic inch to grams per cubic centimeter and compare your results to Jason's.

7. Using your answer from #6 and Jason's original conversion, how does Jason get his final result for how many grams of baking soda he needs?

8. Aside from the punch line here (Why don't you use this half-teaspoon measure instead?), where in the description is an indication that the precision of the conversion factors and results is voided by imprecise measurements? In other words, did Jason estimate at any point, and if so, how could this impact his final answer?

Quantitative Reasoning

Section 4

Linear and Exponential Growth

In Section 4 we study two models for growth: linear growth and exponential growth. Linear growth is characterized by a constant rate of change, and the rate of change of an exponential growth model depends on the quantity that is changing. As often is the case, we encounter other quantitative or mathematical concepts in the following articles that deal with linear and exponential growth. The second article in Section 4 introduces the concept of a weighted average. The content of this section is given below:

- Introduction to interest on money theory.
- Introduction to weighted averages.
- Case Study 4.1: Credit Card Disclosure Statement.
- Case Study 4.2: "Words of a Guru: Math, Plain and Simple" by Amy Rauch-Neilson, *Better Investing,* February 2001.
- Case Study 4.3: "Forcing fuel efficiency on consumers doesn't work" by Jerry Taylor, *Lincoln Journal-Star*, August 21, 2001.

Introduction to Interest on Money Theory

Example. If $500 in an account earns interest at an annual rate of 4%, then the interest earned in one year is 4% of $500 or 0.04 × $500 = $20. At the end of a year, the $20 is added to the $500 so the account now has a balance of $520. Adding 4% of $500 to $500 is the same as multiplying $500 by 1.04 [500 + 500(.04) = 500(1.04)].

The situation above is sometimes referred to as simple interest. That is, there is no compounding, or interest added, during the year. In the above example interest was only added once a year. This is referred to as "compounding annually." Compounding means that periodically, say monthly, the interest earned is added to the account. Let's compute how much money would be in the account after a year if the interest is compounded monthly.

The rate of 4% is an annual rate, so the monthly rate is $\frac{4\%}{12} = 0.33\%$. After a month, the value of the account is $500 + 0.0033($500) = 1.0033 × $500 = $501.65. During the second month, this $501.65 earns 0.33% interest, so at the end of two months the value of the account is 1.0033($501.65) = 1.0033(1.0033)($500) = 1.0033^2($500) = $503.31.

After 3 months the value is 1.0033^3($500) and after 12 months the value is 1.0033^{12}($500) = $520.16. So, compounding monthly results in a gain of 16¢ compared with compounding annually.

If one were compounding daily (say the year has 365 days), the only change is that the daily interest rate of $\frac{4\%}{365} = 0.0109589\%$ is used instead of the monthly interest rate of 0.33%. After a year compounding daily, the value of the account would be $500(1.000109589^{365})$ = $520.40 for a gain of 40¢ over the simple interest method.

Compounding more frequently, say every day, every minute, or every second, would yield more interest. Using the notion of a limit from calculus, one can compute the value of the account after a year of compounding instantaneously (or continuously), that is, interest is added to the account the instant it is generated. Compounding instantaneously the value after a year is $500($e^{0.04}$) = $520.41 very close to the compounding daily amount. (The number e is the base for natural logarithms and is approximately 2.718.)

General Case: Let P be the principal placed in an account at an annual interest rate of r. After t years, compounding n times per year (monthly would make $n = 12$, daily $n = 365$), the amount of money in the account is given by:

$$A = P\left(1 + \frac{r}{n}\right)^{nt}.$$

Compounding continuously yields, $A = Pe^{rt}$.

In both cases r is used in decimal form; that is, if the annual rate is 6%, then use $r = 0.06$ in these formulas.

Note: Here is the limit from calculus that gives the amount of money in an account after t years if the account earns interest at a rate of r annually compounded continuously. Note that r is in decimal form and we are increasing the frequency of compounding to "infinity."

One can compute $\left(1 + \dfrac{1}{x}\right)^x$ for large values of x to see that the values are near 2.718. For example, $\left(1 + \dfrac{1}{1000}\right)^{1000} = 1.001^{1000} = 2.716924$. Then:

$$A = \lim_{n \to \infty} P\left(\left(1 + \frac{r}{n}\right)^n\right)^t = P\left(\lim_{n \to \infty}\left(1 + \frac{1}{\frac{n}{r}}\right)^{\frac{n}{r}}\right)^{rt} = P\left(\lim_{x \to \infty}\left(1 + \frac{1}{x}\right)^x\right)^{rt} = Pe^{rt}.$$

Problem and solution. If \$60,000 is placed in an account earning 5% interest annually, how much is in the account after 10 years (a) if the interest is compounded quarterly, that is, four times per year; (b) if the interest is compounded continuously?

   a. $A = \$60000\left(1 + \dfrac{0.05}{4}\right)^{4 \times 10} = \$60000(1.0125)^{40} = \$98,617.17.$

   b. $A = \$60000 \cdot e^{0.05 \times 10} = \$98,923.28.$

Annual percentage yield. The annual percentage yield (APY) is a percent yield taking into account compounding. For example, if the stated annual interest rate is 6% and interest is compounded monthly, then after one year an investment of \$100 has become:

$$\$100 \times \left(1 + \frac{0.06}{12}\right)^{12} = \$100(1.005)^{12} = \$100(1.0617) = \$106.17.$$

The APY is then 6.17%. This means that compounding monthly at an annual rate of 6% yields the same return as compounding once a year at a rate of 6.17%. (Sometimes the APY is called the effective annual rate.)

More frequent compounding produces a higher APY. The extreme is compounding continuously, which is difficult to visualize physically but completely valid mathematically (as seen above). Compounding continuously at an annual percentage rate of 6% produces $\$100 \times e^{.06} = \$106.18$ for an APY of 6.18%. Consequently, if one invests \$100 for one year at 6% and compounds the interest continuously, one has \$106.18 at year's end, which is the same amount one would have if the \$100 is invested at 6.18% simple interest.

Installment saving. Here we consider building a savings account by periodic deposits. For example, you deposit \$500 on January 1 of each year for 10 years in an account that pays 5% interest compounded annually, with the interest being compounded and added to the account each December 31. How much money is in the account at the end of the 10 years, i.e. on December 31 of the 10th year?

The first $500 earns interest for 10 years, so it amounts to $500 \cdot (1 + .05)^{10} = \$500 \cdot 1.05^{10} = \$814.45$.

The second $500 earns interest for 9 years, so it amounts to $500 \cdot (1 + .05)^{9} = \$500 \cdot 1.05^{9} = \$775.66$.

The third $500 earns interest for 8 years, so it amounts to $500 \cdot 1.05^{8} = \$738.73$.

Continuing in this way until we have the amounts for each of the 10 deposits, the last earning interest for 1 year, so it amounts to $500 \cdot 1.05 = \$525$. The amount of money in the account at the end of 10 years is the sum of these 10 amounts, that is,

$$\$500 \cdot (1.05 + 1.05^{2} + 1.05^{3} + \ldots + 1.05^{10}) = \$6603.39$$

Now, it takes both time and effort to calculate each of the ten amounts and find their sum. The TI calculators have commands that make this easier.

Seq command: Go to LIST and then to OPS. Find "**seq(**" and put this on the home screen. This command will build a sequence for you if you tell it five things: first the formula for the sequence, say $x\text{\textasciicircum}2$; then the variable, in this case x; then the initial value of x, say 3; then the final value of x, say 11; then how much to increase x by each time, say 2. The command looks like this in the calculator:

seq$(x\text{\textasciicircum}2, x, 3, 11, 2)$ and generates $3\text{\textasciicircum}2, 5\text{\textasciicircum}2, 7\text{\textasciicircum}2, 9\text{\textasciicircum}2, 11\text{\textasciicircum}2$ or 9, 25, 49, 81, 121. Try it.

For the purposes of installment savings, we are looking to sum these sequences.

Sum command: Go to LIST and then to MATH. Find "**sum(**" and put it on your home screen. This command will sum a list, or a sequence such as the sequence generated above. Let's try it. If we wanted to both generate and add up the five numbers in the above sequence, we would first put "**sum(**" and follow that by "**seq**$(x\text{\textasciicircum}2, x, 3, 11, 2)$":

sum(seq$(x\text{\textasciicircum}2, x, 3, 11, 2)) = 9 + 25 + 49 + 81 + 121 = 285$.

That one is not difficult by hand. However, we can quickly calculate the sum in the above example of installment savings:

sum(seq$(500*1.05\text{\textasciicircum}x, x, 1, 10, 1)) = \$6603.39$.

While this is on your home screen, hit ENTRY and you will get a copy of this that you can edit. Change the 10 to 20 (or another number of years) and you will get the amount in the account after 20 years (or whatever number of years you enter). After 20 years, the amount is $17,359.63.

Two modifications to note. (1) The **sum(seq(** command above can be altered to the following **sum(seq**$(500*1.05\text{\textasciicircum}x, x, 10, 1, -1)$. This will produce the same sequence to sum, just in a different order. The initial value of the variable is changed from 1 to 10 and the change in x for each step is changed from +1 to –1.

(2) Sometimes one saves up money for, say, a year, and places it in an account at the end of the year. If you alter the situation above to this and ask how much money is in the account at the end of 10 years, the situation is changed in that the first $500 earns interest for 9 years instead of 10 and the 10th year's $500 is placed in the account at the end of the 10 years so has not earned any interest. The command to find this amount would be **sum(seq** $500*1.05\wedge x,x,0,9,1)$ where the final term in the sum is $500*1.05^0 = 500$ because $1.05^0 = 1$.

Exercise: Deposit $20 at the beginning of each month for 10 years in an account earning 6% per year compounded monthly. How much money is in the account at the end of the 10 years?

Note: This appears much more difficult than the example above, but it is not much more difficult once you get the monthly interest rate and the number of months.

Exercise: Suppose you deposit $5 per day every day for 20 years in an account earning 8% interest per year compounded daily. How much will be in the account after 20 years?

This one may overload our calculator since we need to sum up a sequence that has 7300 terms, since there are 7300 days in 20 years. The daily interest rate is $\frac{.08}{365} = .000219178$. So what we want to compute is **sum(seq** $(5*1.000219178\wedge x,x,0,$ $7299,1))$. What one finds is that our calculator is unable to compute this—there are too many terms. So we need to devise another way to calculate. The sequence we are summing is what is called a geometric sequence, where each term is found by multiplying the previous term by a fixed ratio. For example if we call our first term a and the common ratio r, then the sequence is: $a, ar, ar^2, ar^3, \ldots$

What we need is the sum of a certain number of terms of a finite geometric series:

Consider: $\qquad\qquad\qquad\qquad\qquad\qquad S = a + ar + ar^2 + \ldots + ar^n$

Multiply this by r: $\qquad\qquad\qquad\qquad\quad rS = ar + ar^2 + ar^3 + \ldots + ar^{n+1}$

Subtract the second equation from the first: $S - rS = a - ar^{n+1}$

Solve for S: $\qquad\qquad\qquad\qquad\qquad\quad S = \dfrac{a(1 - r^{n+1})}{1 - r}$

Using $a = \$5$, $r = 1.000219178$ and $n = 7299$, we get **sum(seq** $(5*1.000219178\wedge x,x,0,$ $7299,1)) = \dfrac{5(1 - 1.000219178^{7300})}{1 - 1.000219178} = \$90158.71$.

Comment: This formula for the sum of a finite geometric series used above is the way these sums are calculated in practice. For example, if you want to know how much you must save each month for 10 years if you earn 1% per month interest (12% annually) to have $100,000 at the end you use the formula:

$$S = \frac{a(1 - r^{n+1})}{1 - r}$$ with $S = \$100,000$, $n = 120$ months, and $r = 1.01$, and solve for a.

$$100000 = \frac{a(1 - 1.01^{121})}{1 - 1.01}$$

$$\frac{-.01 \times 100000}{1 - 1.01^{121}} = a$$

$$\$428.56 = a$$

In a similar fashion, one can find the amount of an installment payment to pay off a loan over a certain period of time.

Note: The continuous compounding formula is valid for any period t where r is the rate for this period. Thus $A = Pe^{rt}$ becomes $A = Pe^{(0.01)(20)}$, and this gives the amount resulting from investing P dollars at a rate of 1% per month for 20 months.

Exercise: Suppose you deposit $5 each day for 20 years in an account earning 8% per year compounded continuously. How much will be in the account after 20 years?

Here one uses the formula for the amount of money resulting from investing P dollars in an account at an annual rate of r compounded continuously. That amount is given by $A = Pe^{rt}$ where r is expressed in decimal form and t is the number of years. Alternatively, one can use a daily interest rate for r and t as the number of days. The actual annual rate (APY) of interest realized by compounding continuously at an annual rate of r is e^r, so if one compounds interest continuously the annual rate of 8% becomes 8.3287%.

If we deposit the $5 each day and compound continuously at a daily rate of $\frac{8\%}{365} = 0.0219178\%$, the amount of money after 20 years is given by the following sum:

$$\$5(1 + e^{.000219178} + e^{2(.000219178)} + \ldots + e^{7299(.000219178)})$$

Or, after computing $e^{.000219178} = 1.0002192$,
$$\$5(1 + 1.0002192 + 1.0002192^2 + 1.0002192^3 + \ldots + 1.0002192^{7299})$$

This sum has too many terms for most calculators, so the **sum(seq(** method will not produce an answer. However, the formula for the sum of this geometric series will produce an answer as follows:

$$\frac{\$5(1 - 1.0002192^{7300})}{1 - 1.0002192} = \$90,167.80$$

Note that this is only about $11 more than if one compounded daily.

Introduction to Weighted Averages

A **weighted average** is an average that takes into account the proportional relevance of each component, rather than treating each component equally. In general, the weighted average of n quantities $a_1, \ldots, a_n$ whose weights are $w_1, \ldots, w_n$, respectively, is:

$$\frac{w_1 a_1 + w_2 a_2 + \ldots + w_n a_n}{w_1 + w_2 + \ldots + w_n}.$$

The "normal average," or arithmetic mean, of the quantities $a_1, \ldots, a_n$ is just the weighted average with all weights of value 1, or $\dfrac{a_1 + a_2 + \ldots + a_n}{1 + 1 + \ldots + 1} = \dfrac{a_1 + a_2 + \ldots + a_n}{n}$.

Example 1. A student has four test scores (out of 100) of 75, 85, 87, and 70. The scores are worth different amounts in determining the final average; the first is 20%, the second is 10%, the third is 30%, and the fourth is 40%. The final average is:

$$\frac{0.20 \times 75 + 0.10 \times 85 + 0.30 \times 87 + 0.40 \times 70}{0.20 + 0.10 + 0.30 + 0.40} = \frac{77.6}{1} = 77.6$$

or

$$\frac{20 \times 75 + 10 \times 85 + 30 \times 87 + 40 \times 70}{20 + 10 + 30 + 40} = \frac{7760}{100} = 77.6.$$

Note that the non-weighted average of the four test scores would be given by

$$\frac{75 + 85 + 87 + 70}{4} = 79.25.$$

Example 2. A customer buys apples of five different varieties at a supermarket. The amounts and costs are below.

Variety	Number of pounds purchased	Cost per pound
Red Delicious	3	$1.99
Yellow Delicious	4	$2.29
Granny Smith	2.5	$1.89
Fuji	5	$2.69
Jonagold	1.5	$2.09

What is the average cost of a pound of apples purchased?

$$\frac{3 \times \$1.99 + 4 \times \$2.29 + 2.5 \times \$1.89 + 5 \times \$2.69 + 1.5 \times \$2.09}{3 + 4 + 2.5 + 5 + 1.5} = \$2.28.$$

Notice that this *average cost of a pound of apples purchased* is different from the *average price per pound of the five varieties*, which is:

$$\frac{\$1.99 + \$2.29 + \$1.89 + \$2.69 + \$2.09}{5} = \$2.19.$$

Exercise 3. The shooting performance of two basketball players, Tayshaun and Corliss, is given below for two halves of a game.

Player	1st Half			2nd Half		
	Baskets	**Attempts**	**Percent**	**Baskets**	**Attempts**	**Percent**
Tayshaun	4	11	36.4%	3	4	75%
Corliss	1	5	20%	7	10	70%

1. For the game, which player had the best shooting performance? Compute the game percents. (This is an example of what is called Simpson's Paradox [so-called because Edward Simpson noted it in 1951].) The paradox is this: *Better in each case but worse overall.*

2. Obviously, sometimes better in each case yields better overall. Can you explain what there is about the above example that creates Simpson's Paradox?

The answer can be seen from a weighted average:

$$\text{Tayshaun's average: } \frac{11 \times .364 + 4 \times .75}{15} = 0.467$$

$$\text{Corliss' average: } \frac{5 \times .2 + 10 \times .7}{15} = 0.533$$

Notice that Tayshaun's worst percentage (36.4%) is multiplied by his larger weight (number of attempts) while Corliss' best percentage (70%) is multiplied by his smaller weight.

Exercise 4. Here is an example of Simpson's Paradox using test scores of four sub-populations of students—A, B, C, and D, which together make up the entire student population. The following table gives the number of students in the four groups testing at the "proficient" level in two different districts. Which district has the best record of students testing "proficient"? Does this illustrate Simpson's Paradox?

	District A			District B		
	Total	**Proficient**	**%**	**Total**	**Proficient**	**%**
A	4000	3000		1000	760	
B	200	100		900	500	
C	400	200		150	80	
D	400	300		400	320	
Total	5000	3600		2450	1660	

Set up the weighted average that explains this occurrence of Simpson's Paradox.

Exercise 5. Most colleges and universities compute students' grade point averages (GPAs) as a summary measure (condensed measure) of the students' academic records. Compute the grade average of a student who has the following grades: A in a 3-hour sociology course; B in a 4-hour chemistry course; C in a 2-hour fine arts course; and B in a 3-hour mathematics course. What are the weights? What does the number you compute mean?

Exercise 6. In baseball, slugging percentage (SLG) is calculated by crediting a player one point for each single (1B), two points for each double (2B), three points for each triple (3B), and four points for each home run (HR), and then dividing the sum of points awarded by the player's times at bat (AB). Compute the slugging percentage for each of the players listed below. What does the number you compute mean?

	AB	1B	2B	3B	HR	SLG
Rodriguez	583	98	31	0	54	
Pujols	565	114	38	1	32	

Case Study 4.1: Credit Card Payments

Resource Material: Individually obtained Credit Card Disclosure Statement from a variety of possible sources. For example: www.capitalone.com.

Learning Goals: The learning goals of this case study include careful reading of financial rules and regulations, working with interest, performing calculations with recursive formulas, and calculating a revolving monthly balance on a credit card statement. Terms such as APR, grace period, and minimum payment are introduced and used.

This case study addresses the basic issue of credit card debt and what happens if the card holder makes only the minimum payments on the credit card balances. In addition, these exercises explore what occurs if the card holder misses one or two payments.

Warm Up Exercises for Case Study 4.1

1. An annual interest rate of 13.5% would be equivalent to what monthly rate? What would be the daily rate?

2. What annual percentage rate (APR) would incur a monthly interest charge of $14 on a balance of $1,700?

3. If you paid $27 on your last month's credit card balance of $950 and your credit card has an advertised APR of 13.5%, what is the balance on this month's credit card bill?

4. Jackson is paying his current credit card bill. His bill says he had a starting balance of $2,500, an interest charge of $31.25, and a credit of $45 (from his last payment).

   a. What is the remaining balance on his credit card?

   b. His credit card company requires him to pay a minimum payment that consists of the current interest charge plus 1% of his balance. What is Jackson's minimum payment this month?

   c. If Jackson makes only this minimum payment, how much of his payment goes towards interest and how much goes towards paying off the balance on the credit card? *Give you answer in both dollars and percents.*

   d. What is the remaining balance on his credit card the next month?

Article for Case Study 4.1

[By checking your mailbox or visiting various financial websites, locate a current credit card offer. Be sure to retrieve the full disclosure of all the rules and regulations pertaining to the credit card.]

Study Questions for Case Study 4.1

1. **Obtain a current credit card offer and read through the terms. Identify the following information. (You may have to dig into the "fine print."):**

   a. **APR (Annual Percentage Rate): Note that there are often different categories (payment, transfer) and rates (introductory and regular). These rates can also be fixed or variable. Determine the regular payment APR for your card. If the rate is listed as variable, indicate how this rate is determined. Does your card have a Default APR? If so, what is it and when does it get used?**

   b. **Grace Period: Most credit cards have a grace period on new purchases. State whether your card has a grace period and when it applies.**

   c. **Does your card have an annual fee? If so, what is it?**

   d. **How is the minimum payment determined?**

   e. **What happens if you miss one or more payments?**

2. **Credit cards allow the holder to purchase an item and take it home to enjoy while paying for the item at a later date. Purchasers may pay for the item in full when they receive their first statement. For companies that offer "grace periods," no additional interest is charged. However, often people use credit cards when they plan to pay for an item over several months or years. If one does not pay off the entire credit card balance, interest will be charged on ALL purchases, and the interest owed is then calculated from the date of purchase (not the date the bill is received). Thus, there is no grace period at all.**

   Suppose you have your eye on a nice big screen TV that costs $2,000. You decide to use your new card to make the purchase. Assume that you do not plan to pay the debt off in full at the first billing, so you will be subjected to interest charges each month.

   a. **Using the regular purchase APR, determine the monthly interest rate for your card.**

   b. **If your card has a grace period, when you receive your first bill (say, in one month), there should be no interest charge. If you make the required minimum payment, how much will you have to pay the first month? What is the remaining balance (amount you owe) on the card?**

   c. **Assuming you only make the required minimum payment, your second bill will have an interest charge. Assuming you are charged one month of** interest on the remaining balance, how much will you be charged in interest on your second bill?

d. Your new minimum payment will be computed by using the starting balance plus the interest charged. What is the minimum payment required on your second bill?

e. Assuming that you again just make the minimum payment, what percent of your payment goes towards interest and what percent goes towards reducing your original balance of $2,000?

f. When your third bill comes, what is the new balance and minimum payment?

3. Suppose you keep making the minimum payment for one year.

a. After one year, what is the current balance on the credit card?

b. After one year, how much have you paid the credit card company?

c. Of all the money you paid the credit card company, what percent went towards reducing the original balance of $2,000, and what percent went towards interest payments?

d. If you keep making minimum payments, how long do you think it will take to pay off the purchase of the TV? *An estimate based on your work so far will suffice.*

4. Use a spreadsheet, such as Excel, to keep track of the following quantities:

a. Statement number (first, second, etc.)

b. Interest charged that month

c. Current balance

d. Minimum payment required

e. Total amount of interest paid to date

f. Total of all payments made

g. Percent of payment which is interest

How long will it take to pay off the original purchase price of $2,000? How much interest would you have to pay for this privilege?

5. How would the scenario change if you could afford to pay twice the minimum payment every month?

6. What happens if, for some reason, you miss your payments twice during the first year?

7. Is it worth "shopping around" for a credit card with a lower APR? Explain when this would be advantageous and why.

Case Study 4.2: Math, Plain and Simple

Resource Material: "Words of Guru: Math, Plain and Simple" by Amy Rauch-Neilson *Better Investing*, February 2001.

Learning Goals: The learning goals of this case study include computing simple and compound interest, computing the amount of money resulting from installment savings with compounding interest, computing weighted averages, and analyzing claims about the results of installment savings and weighted averages.

The article in this case study addresses two areas of quantitative reasoning: installment savings and weighted averages. The installment savings example stems from a statement that saving $2.50 per day for 25 years with compounding interest can produce a substantial amount of money. The weighted averages example concerns the average cost of a share of stock if one purchases shares by a method called dollar-cost averaging.

Warm Up Exercises for Case Study 4.2—Installment Savings

1. Assume that $12,000 is placed in a savings account that pays 6.4% interest annually.

    a. Find the amount of money in the account after 8 years if the interest is compounded annually.

    b. Find the amount of money in the account after 8 years if the interest is compounded quarterly (every three months).

    c. Find the amount of money in the account after 8 years if the interest is compounded monthly.

2. On January 1 of each year for 8 years you deposit $3,600 in an account that earns 7% interest per year with the interest being computed and added every December 31 (compounding annually).

    a. Write out the sum that gives the amount of money in the account on December 31 of the 8th year.

    b. Use the **sum** and **seq** commands on your calculator and compute the sum in part (a). Give the command that you enter into your calculator.

3. On January 1 you begin saving money at $300 per month. At the end of each year (December 31) for 8 years you place the savings in an account earning 7% interest per year compounded annually.

   a. Write out the sum that gives the amount in the account on December 31 of the eighth year.

   b. Using the **sum** and **seq** commands on your calculator, find the sum in part (a).

   c. How does the command you enter into your calculator here differ from what you entered in #2 above? Explain why there are differences.

Warm Up Exercises for Case Study 4.2—Weighted Averages

A **weighted average** is an average that takes into account the proportional relevance of each component, rather than treating each component equally.

1. If oranges cost $0.30, apples $0.25, and bananas $0.20 each, find the average cost of the fruit in a basket that has 11 oranges, 15 apples, and 22 bananas. Is this different from the average of $0.30, $0.25, and $0.20? Explain why.

2. Student grades in a course are determined by scores (out of 100) on four examinations, call them 1, 2, 3, and 4. Exam 1 constitutes 15% of the grade, Exam 2 is 20%, Exam 3 is 30%, and Exam 4 is 35%. What is a student's average who has the following scores? 1—60, 2—65, 3—75, and 4—90. Compute both the weighted average and the average (also called the arithmetic mean).

3. Assume that over a period of time you purchase shares of a fund according to the following investment schedule.

Amount invested	$500	$800	$400	$1200	$700	$1200	$400	$800
Price per share	$10	$12	$13	$11	$13	$15	$16	$15
# Shares purchased								

   a. Complete the above table and find the total number of shares purchased. (One can purchase fractions of shares of the fund.)

   b. Find the average cost of the shares you purchased.

   c. Find the average of the prices per share at the times the investments were made.

   d. Explain why the answers to (b) and (c) differ.

Article for Case Study 4.2

Better Investing
February 2001
Math, Plain and Simple
Words of a Guru
By AMY RAUCH NEILSON

Q. If a stock splits, as an investor, have I doubled my money?
A. No. This is one of the most commonly held misconceptions of beginning investors.
You haven't doubled your money. What's doubled, at least in a two-for-one split, is
the number of shares that you now own.

You can use a simple math equation to figure out how many shares of stock you own
following a split: Take the number of shares you own and multiply them by the fraction
created by the split. For example, in a 2-for-1 split, that fraction would be 2/1. So, if you
own 500 shares, then your equation will be: $500 \times 2/1 = 1,000$ total shares.

In most cases, a stock split is a sign of a growing, healthy company. It also places the
market price in a range that is more accessible to many investors, particularly individuals.
Splits give long-term investors the opportunity to acquire a greater number of shares and
build wealth.

Is successful investing really just a matter of common sense?

It's been nearly a decade since investment guru Peter Lynch's first book, Beating the
Street, hit the bookstores. It rose quickly on the best-seller lists not only on sales to readers who were investors, but perhaps more importantly, to readers who weren't.

The popularity of Lynch and his investment philosophy is the result of a powerful combination. Lynch has not only made himself oodles and oodles of money by applying his principles, but, as manager of the Fidelity Magellan Mutual Fund from 1977–1990, he literally
shared the wealth with thousands of Fidelity Magellan investors. During those 13 years
that he was the fund's manager, the fund enjoyed a whopping 2,700 percent return.

Could it really be so simple to be a stock market success? The answer, not just from
Peter Lynch, but from the hundreds of thousands of NAIC members across the globe, is
a resounding "yes." In fact, Lynch's philosophy walks hand-in-hand with NAIC's own,
notes Billy Williams, chairman and past president of the NAIC Atlanta Chapter.

Lynch has authored three books, more than 50 articles and one CD-ROM in his continuing quest to educate beginning investors. One of his most recent efforts to educate

investors is his "Key Things Every Investor Should Know," located at www400.fidelity. com. This is a guided tour through investment basics like stocks, bonds, and mutual funds as well as Lynch's insights on why you should own a home and pay off any high-interest debt before you begin investing.

Once you've spent the 30 to 45 minutes it takes to make your way through the guide, you can go back to the home page and test your knowledge by taking the Invest Test. Don't blow past this opportunity because you consider yourself to be an "experienced" investor; Lynch contends there's plenty there for you, too. A sampling of the content follows.

The fate of your portfolio rests in your hands

"Why must you learn about investing?" Lynch asks near the beginning of the guide. "You can hire a financial planner or manager to make the decisions, cash your dividend checks, and try to forget about the whole business. But no matter how much responsibility you delegate, and how much help you receive, the fate of your portfolio rests in your hands. Think of it this way: you're the CEO of an important enterprise called 'Your Financial Future.'"

It's a theme Lynch has promoted for years: "Know what you own and know why you own it." Why, you ask? "Because you should know how to allocate your assets among stocks, bonds and cash, so you're comfortable with the risk and satisfied with the potential reward," he says. "You should be able to describe the types of investments you own (growth, value, large-cap, small-cap, high yield bonds, government bonds, etc.) and how they are expected to perform relative to other types of investments."

The payoffs from petty thrift can be remarkable

"This is the point I try to make to anyone who tells me that he or she can't possibly squeeze a dime out of the household budget now to invest in the future." Lynch's example really hits home: "Forego the daily cup of $2.50 cappuccino and invest the proceeds. In 25 years, assuming you can orchestrate a mediocre eight percent annual rate of return, you'll have $72,800 before taxes to show for your caffeine sacrifice. If the price of coffee goes up, as it surely will, you'll have a bigger payoff."

Dollar-cost averaging

This is the way hundreds of thousands of NAIC investors have built large portfolios with small sums of money over a period of time. It's called dollar-cost averaging.

"There's a clever way to take the worry out of buying stocks or stock mutual funds. It's called 'dollar-cost averaging,'" Lynch says. "Here's how it works. Suppose you have $3,000 you're planning to invest in a growth fund. Instead of making a single lump-sum payment, you decide to spread your $3,000 over a one-year period, investing $250 per month.

"In January, when the share price is 10, your $250 buys 25 shares. By October, the price has fallen to 7. Now your $250 buys 35.71 shares. In November, the price rises again, and by December it's back to its January level of $10.

"The price started the year at 10 and finished the year at 10, and fluctuated from a high of 15 to a low of 7 in between. By spreading out your investment, you bought more shares when the price was low, and fewer shares when the price was high. The average price you paid was less than 10. In fact, it was 9.93.

"That's the important feature of dollar-cost averaging. In up-and-down markets, it can reduce your cost of investing. The bottom line in this example? By paying 9.93 per share instead of the average price of 10.42 per share, you saved 5 percent.

"Gyrations in price don't bother the dollar-cost averager. You're buying stocks or mutual funds at cheap rates when the market falls and expensive rates when the market rises, but you'll never invest a lump sum at the worst possible moment, as investors did in September 1929 and August 1987. Dollar-cost averaging helps you build a portfolio in a systematic, cost-effective way."

Study Questions for Case Study 4.2

"Math, Plain and Simple, Words of a Guru" by Amy Rauch Neilson
Better Investing
February 2001

A. Quoting Peter Lynch, the final paragraph of Page 1 is:

The payoffs from petty thrift can be remarkable

"This is one point I try to make to anyone who tells me that he or she can't possibly squeeze a dime out of the household budget now to invest in the future." Lynch's example really hits home: "Forego the daily cup of $2.50 cappuccino and invest the proceeds. In 25 years, assuming you can orchestrate a mediocre eight percent annual rate of return, you'll have $72,800 before taxes to show for your caffeine sacrifice. If the price of coffee goes up, as it surely will, you'll have a bigger payoff."

1. Assuming no interest earnings, how much money is saved in 25 years doing as the writer suggests?

2. If the cappuccino savings are collected for the first year and then placed in a savings account earning 8% interest, compounded annually, how much interest is earned on the first year's savings during the second year? Find the amount of savings at the end of year 2.

3. Extending the process began in #2, find the amount of savings at the end of three years if the money is saved each year and then placed in the savings account earning 8% compounded annually.

4. Continuing the process as in #3, find the savings at the end of five years.

5. Extend the reasoning in #4 to find the amount of savings after 25 years if the annual savings, at the end of the year, is placed in the account earning 8% compounded annually. Compare your results to that claimed by Lynch. (You will probably need to use something like the sum and seq commands on the TI calculators.)

6. Repeat the calculation in #5 above compounding monthly instead of annually. Compare your results to that claimed by Lynch.

7. Repeat the calculations from #6 compounding interest daily. Compare your result to that claimed by Lynch.

8. Assume that the $2.50 per day is placed in an account at the end of the day and earns interest compounded continuously for the remainder of the 25 years. Find the amount of money in the account at the end of the 25 years.

B. Dollar-cost averaging

From page 2, paragraph 2:

"Here's how it works. Suppose you have $3000 you are planning to invest in a growth fund. Instead of making a single lump-sum payment, you decide to spread your $3000 over a one-year period, investing $250 per month.

"In January, when the share price is 10, your $250 buys 25 shares. By October, the price has fallen to 7. Now your $250 buys 35.71 shares. In November the price rises again, and by December it's back to the January level of $10.

"The price started the year at 10 and finished the year at 10, and fluctuated from a high of 15 to a low of 7 in between. By spreading out your investment, you bought more shares when the price was low, and fewer shares when the price was high. The average price you paid was less than 10. In fact it was $9.93.

"That's the important feature of dollar-cost averaging. In up-and-down markets, it can reduce your cost of investing. The bottom line in this example? By paying $9.93 per share instead of the average price of $10.42 per share, you saved 5 percent.

"Gyrations in price don't bother the dollar-cost averager. . . . you'll never invest a lump sum at the worst possible moment, as investors did in September 1929 and August 1987."

1. From the information given, what do you know about the price per share of this fund in each of the months of the year?

2. Explain why the following two examples of monthly price per share of this fund satisfy the conditions in #1. Fill in the number of shares purchased each month and compute the average cost of a share for each of the two examples.

	Jan	Feb	Mar	Apr	May	Jun	Jul	Aug	Sep	Oct	Nov	Dec
A	$10	$15	$15	$15	$15	$15	$15	$15	$15	$7	$15	$10
Shares	25											25

	Jan	Feb	Mar	Apr	May	Jun	Jul	Aug	Sep	Oct	Nov	Dec
B	$10	$7	$7	$7	$7	$7	$7	$7	$7	$7	$15	$10
Shares	25											25

3. In view of the results of #2 above, critique the statement by the writer: "The average price you paid was less than 10. In fact, it was $9.93." If you follow the advice on dollar-cost averaging as described in this article will the average price you pay per share always be less than $10?

4. Critique the statement by the writer: "By paying $9.93 per share instead of the average price of $10.42 per share, you saved 5 percent."

5. Critique the statement: [Using dollar-cost averaging] "you'll never invest a lump sum at the worst possible moment."

Case Study 4.3: Forcing Fuel Efficiency on Consumers Doesn't Work

> Resource Material: "Forcing fuel efficiency on consumers doesn't work" by Jerry Taylor, Cato Institute, *Lincoln Journal-Star*, August 21, 2001.

Learning Goals: The learning goals of this case study include analyzing quantitative arguments about economics of fuel efficiency, devising reasonable assumptions for information not supplied by the writer, analyzing the effects of different assumptions on costs and savings, using linear and exponential equations to model costs and savings, graphing linear and exponential equations, using graphs to compare costs and savings over time, and interpreting features of graphs in terms of the situation being modeled.

This Op-Ed article argues that fuel efficiency resulting from government requirements on automobiles and trucks is not economically sound. There are several quantitative assertions in the article, some of which can be critiqued after making assumptions about various quantities that are not given in the article.

The mathematical concepts that are involved in critiquing the article's assertions include linear equations (sometimes specialized and called cost equations) and exponential equations that give the amount of money in accounts earning interest.

Warm Up Exercises for Case Study 4.3

1. Assume the gasoline version of a Honda Civic costs $21,000 and gets 35 MPG while the hybrid version costs $26,500 and gets 47 MPG. Assume you drive 11,000 miles per year and gasoline costs $2.90 per gallon. Let G be the purchase cost plus the gasoline cost for x years of the gasoline versions and H be the purchase cost plus the gasoline cost for x years for the hybrid version of the Civic.

   a. Give G and H as equations in terms of x.

   b. Give the graphs of the equations from (a) on the rectangular coordinate axes system using a horizontal axis scaled in years from 0 to 50 and a vertical axis scaled in dollars from $20,000 to $50,000.

   c. How many years will be required to break even on the extra cost of the hybrid? What part of the graph exhibits this answer?

2. Assume the yearly savings in gasoline costs due to the purchase of a hybrid automobile are $400 and the extra cost of purchasing the hybrid is $4500.

   a. Produce an equation of net savings over a period of x years.

   b. Graph this equation on an appropriate rectangular coordinate axis system and identify the point that indicates the number of years required to break even. How long does it take to break even?

Article for Case Study 4.3

Lincoln Journal-Star
August 21, 2001
Forcing fuel efficiency on consumers doesn't work
By JERRY TAYLOR

Although the late, great energy crisis seems to have come and gone, the political fight over yesterday's panic rages on. The big dust-up this fall will be over SUVs, light trucks and minivans. Should the government order Detroit to make them get more miles per gallon? Conservationists say "yes." Economics 101 says "no."

Let's start with a simple question: Why should the government mandate conservation? When fuel becomes scarce, fuel prices go up. When fuel prices go up, people buy less fuel. Economists have discovered over the long run, a 20 percent increase in gasoline costs, for instance, will result in a 20 percent decline in gasoline consumption. No federal tax, mandate or regulatory order is necessary.

Notice the phrase "over the long run." Energy markets are volatile because consumers do not change their buying habits much in the short run.

This has led some critics to conclude that people don't conserve enough when left to their own devices. They do, but consumers need to be convinced that the price increases are real and likely to linger before they'll invest in energy-efficient products or adopt lifestyle changes. But even in the short run, people respond. Last summer was a perfect example: For the first time in a non-recession year, gasoline sales declined in absolute terms in response to the $2 per gallon that sold throughout much of the nation.

Mandated increases in the fuel efficiency of light trucks, moreover, won't save consumers money. A recent report from the National Academy of Sciences, for instance, notes that the fuel efficiency of a large pickup could be increased from 18.1 miles per gallon to 26.7 miles per gallon at a cost to automakers of $1,466. But do the math: It would take the typical driver 14 years before he would save enough in gasoline costs to pay for the mandated up-front expenditure. A similar calculation for getting a large SUV up to 25.1 miles per gallon leads to a $1,348 expenditure and, similarly, more than a decade before buyers would break even.

"Fine with me," you say? But it's one thing to waste your own money on a poor investment; it's entirely another to force your neighbor to do so. You could take that $1,466, for instance, put it in a checking account yielding 5 percent interest, and make a heck of a lot more money than you could by investing it in automobile fuel efficiency.

Even if government promotion of conservation were a worthwhile idea, a fuel efficiency mandate would be wrong. That's because increasing the mileage a vehicle gets from a gallon of gasoline reduces the cost of driving. The result? People drive more.

Reprinted with permission from *Lincoln Journal-Star,* August 21, 2001.

Energy economists who've studied the relationship between automobile fuel efficiency standards and driving habits conclude that such mandates are offset by increases in vehicle miles traveled.

If we're determined to reduce gasoline consumption dramatically, the right way to go would be to increase the marginal costs of driving by increasing the tax on gasoline. Now, truth be told, I don't support this idea much either. A recent study by Harvard economist Kip Viscusi demonstrates that the massive fuel taxes already levied on drivers (about 40 cents per gallon) fully "internalize" the environmental damages caused by driving. But conservationists reject this approach for a different reason: Consumers hate gasoline taxes and no Congress or state legislature could possibly increase them.

Look, it's a free country. If you want to buy a fuel-efficient car, knock yourself out. But using the brute force of the government to punish consumers who don't share your taste in automobiles serves no economic or environmental purpose.

Study Questions for Case Study 4.3

"Forcing fuel efficiency on consumers doesn't work" by Jerry Taylor
Lincoln Journal-Star
August 21, 2001

The following quantitative assertions are made in the article.

I.   Economists have discovered that, over the long run, a 20 percent increase in gasoline costs, for instance, will result in a 20 percent decline in gasoline consumption.

II. A recent report from the National Academy of Sciences, for instance, notes that the fuel efficiency of a large pickup could be increased from 18.1 miles per gallon to 26.7 miles per gallon at a cost to automakers of $1,466.

III. But do the math: It would take the typical driver 14 years before he would save enough in gasoline costs to pay for the mandated up-front expenditure [$1466].

IV. A similar calculation for getting a large SUV up to 25.1 miles per gallon leads to a $1,348 expenditure and, similarly, more than a decade before buyers would break even.

V.   You could take that $1,466, for instance, put it in a checking account yielding 5 percent interest, and make a heck of a lot more money than you could by investing it in automobile fuel efficiency.

1. **Answer the following:**

   a. **How could you check assertions I and II?**

   b. **What assumptions would need to be made in checking assertion III?**

   c. **What assumptions would need to be made in checking assertion IV?**

   d. **What assumptions would need to be made in checking assertion V?**

2. **Answer the following:**

   a. **Is the assertion III above reasonable? Explain why or why not.**

   b. **What would be the effect of increased costs of gasoline on assertion III? Justify your answer with quantitative evidence. For example, illustrate this effect with an example of increased costs.**

   c. **What would be the effect of increased miles driven per year on assertion III? Justify your answer with quantitative evidence.**

   d. **Assume the cost of gasoline in 2001 was $1.40 per gallon and that it would take 14 years for the "typical driver" to recover the $1466 through savings in gasoline costs. How many miles per year would the "typical driver" drive?**

3. Answer the following:

a. Is the assertion IV above reasonable? Why or why not?

b. How would the savings be affected if the current MPG of large SUVs were lower than 18.1 MPG? Illustrate this with an example.

4. Answer the following:

a. Is assertion V above reasonable? Why or why not? You should include a graph of the amount of bank savings and the amount of gasoline savings for 40 years.

b. If the $1466 is placed in one account at 5% interest and the annual savings from gasoline are deposited in a second account earning 5% interest, compounded annually, how do the amounts in the two accounts compare? You should include a graph of the bank savings and the amount of gasoline savings for 15 years.

5. Use $1.40 per gallon for gasoline and 12,000 miles per year and compare the amount of money in a bank account where the $1,466 is placed at 5% interest, to the savings in gasoline with no interest earned on the savings over x years. You should include a graph of the amount of bank savings and the amount of gasoline savings for 45 years.

Quantitative Reasoning

Section 5

Graphical Interpretation and Production

In Section 5 we study graphs, which are simply visual representations of quantitative information. We will interpret, critique, and produce a variety of graphs and formalize an understanding of how these objects are used to convey information. The content of this section is given below.

- Case Study 5.1: "Enrollment Rates Rise" by Jeff Smith, *Morning News of Northwest Arkansas*, September 18, 2004, and "UA Enrollment up 5% over 2003" by Chris Branam, *Arkansas Democrat-Gazette,* September 18, 2004.
- Case Study 5.2: "Graphic entitled Number of Students on Central Plaza on a Typical Week Day," unknown origin.
- Case Study 5.3: "Two Views of a Tax Cut graphic," *New York Times*, April 17, 1995.
- Case Study 5.4: "Sexual Betrayal" graphic, *Journal of Youth and Adolescence* 29:4 (2000).
- Case Study 5.5: "Decade After Health Care Crisis, Soaring Costs Bring New Strains" by Robin Toner and Sheryl Gay Stolberg, *New York Times*, August 11, 2002.

Case Study 5.1: Enrollment Rises

Resource Material: "Enrollment Rates Rise" by Jeff Smith, *The Morning News of Northwest Arkansas*, September 18, 2004, and "UA Enrollment up 5% over 2003" by Chris Branam, *Arkansas Democrat-Gazette*, September 18, 2004.

Learning Goals: The learning goals of this case study include comparison and analyses of graphing techniques and distinguishing between a quantity and a rate of change of that quantity.

This case study focuses on two newspaper reports of the same event, the release of the fall enrollment numbers by the University of Arkansas. The two articles use different types of graphs to represent the fall enrollments 1997 to 2004. The two reports of enrollments agree except for the fall 1998 enrollment, and only one of the graphs gives the 1997 enrollment. Determining which 1998 number is correct is part of the study.

One of the two articles also reports graduation rates. This is reflected in the article's headline "Enrollment Rates Rise" and the sub-heading, "More Students Graduate." Analyzing whether these two headlines are supported by the story is part of the study.

Warm Up Exercises for Case Study 5.1

Johan's hourly wages for 2000–2008 are given in the table below.

year	2000	2001	2002	2003	2004	2005	2006	2007	2008
wage	$6.00	$6.00	$6.50	$7.00	$7.50	$8.00	$8.00	$8.50	$9.00

1. Graph these data using a bar graph.

2. Graph these data using a line graph with the scale on the vertical axis going from $6 to $9 graduated in increments of $0.50.

3. Graph these data using a line graph with the scale on the vertical axis going from $0 to $9 graduated in increments of $0.50.

4. Which graph offers the best visual representation of the data? Why?

Articles for Case Study 5.1

The Morning News of Northwest Arkansas
September 18, 2004
Enrollment Rates Rise
MORE STUDENTS GRADUATE
By JEFF SMITH

FAYETTEVILLE—The University of Arkansas posted another record fall enrollment with 17,269 students.

UA officials also are celebrating another all-time high: the sixth-year graduation rate hit 52.8 percent.

Both enrollment and graduation rates have steadily increased since John White became Chancellor in 1997. UA had 14,740 students and a graduation rate of 41.8 percent that year.

But White said Friday he is not satisfied. He still plans to reach 2010 commission goals of 22,500 students and a graduation rate of 66 percent.

UA Enrollment

The University of Arkansas continues to set record enrollment numbers with 17,269 students on campus this fall, representing a 4.9 percent increase from last year.

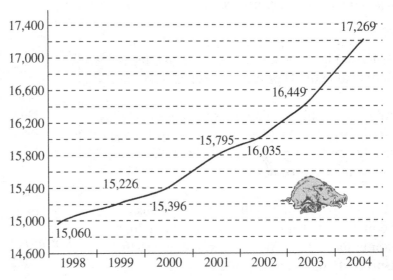

SOURCE: University of Arkansas

GABRIEL CALZADA • THE MORNING NEWS

Arkansas' graduation rate ranks it last in the 54 schools in the major NCAA athletic conferences.

More than 550 entering freshmen in 1998 received Chancellor's Scholarships, which at the time covered tuition and fees. The previous year, slightly more than 50 were awarded.

The University currently gives $8,000-a-year scholarships to about 250 students a year.

"1998 has transformed the University of Arkansas," White said, adding, "each class that comes after that is going to be influenced by (that class)."

Johnetta Cross Brazzell, vice chancellor for student affairs, said the university has created an inviting environment where students want to stay, excel, and graduate.

She said that the increased graduation rate is "predictable given the programs we've put in place and given the quality of students we're bringing here."

The 2004 freshman class of 2,501 students is the second largest, 51 smaller than the 1998 class.

Dawn Medley, UA director of admissions, said her office specifically targeted neighboring states, whose students who excel academically can receive a waiver of out-of-state tuition.

She said recruiters discuss the UA's growing national reputation, its ranking as the best college buy and the importance of an education.

"We were able to go out to speak to the students and let them know what a great educational value the U of A had to offer," she said.

Admission officers plan to expand in the Dallas area as well as in Oklahoma and central Arkansas by conducting more high school visits and financial aid workshops.

"It's not hard to sell in the state of Arkansas. People know that if you want a quality education in the state of Arkansas, you come to the University of Arkansas," Medley said. "What's really nice is that people outside the state are starting to realize that as well."

FAST FACTS
UA GRADUATION RATES
The University of Arkansas has steadily improved its six-year graduation rate. For instance, 52.8 percent of the students enrolling as freshmen in Fall 1998 graduated by Spring 2004.

- 1998—52.8 percent
- 1997—48.1 percent
- 1996—45.9 percent
- 1995—44.8 percent
- 1994—45.3 percent
- 1993—45.1 percent
- 1992—43.5 percent
- 1991—41.8 percent

SOURCE: University of Arkansas

The University saw gains in minorities, a critical area identified by UA officials and state lawmakers. The Native American freshmen enrollment almost doubled from 32 to 62, and 55 Hispanics enrolled this year compared to 38 last year.

African-American enrollment grew from 111 last year to 115, a 3.6 percent increase, and 71 Asian freshmen are on campus, up from 66 last year. Freshmen international student enrollment saw a 23 percent decline, from 42 last year to 32.

A complete breakdown of univesitywide minority enrollment wasn't available Friday. UA officials said they haven't fully analyzed all student data because they are changing computer software systems.

The university offered 66 students $5,000 to $8,000 annual scholarships beginning this fall aimed at improving diversity. The Silas Hunt Scholarships were announced last fall as UA administrators faced a state legislative meeting about declining minority enrollment.

Brazzell said the university is working intensely to increase the number of minority students attending UA.

She said officials are pleased with current results but aren't satisfied.

"We're not where we are going to be," Brazzell said.

Freshman ACT scores remained the same from last year at 25.4, while the average high school grade point average dropped from 3.6 to 3.57 this fall.

Arkansas Democrat-Gazette
September 18, 2004
UA enrollment up 5% over 2003
By CHRIS BRANAM

The University of Arkansas at Fayetteville has recorded its highest annual increase in fall enrollment since Chancellor John A. White took over in 1997.

UA enrollment

■ Fall enrollment at the University of Arkansas at Fayetteville, recorded Sept. 7, grew by 5 percent over the previous year, the university announced Friday. UA's fall enrollment has increased by 20 percent since Chancellor John A. White joined UA in 1997. White has set an enrollment goal of 22,500 by 2010 for the state's largest university. Here is a breakdown of UA's year-by-year enrollment during White's tenure:

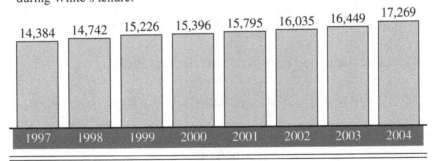

SOURCE: University of Arkansas at Fayetteville   Arkansas Democrat-Gazette

The campus has 17,269 students as of the official Sept. 7 count, the university announced Friday.

That's a 5 percent increase from the previous year.

"The new numbers represent a lot of hard work on the part of staff and faculty, and, of course, our students," White said in a statement. "They are remarkable accomplishments for both the university and the entire state."

The previous best increase under White was 3.3 percent in 1999. This is the fourth consecutive fall that the 133-year-old land-grant campus has seen record enrollment.

"It's wonderful news," said UA Admissions Director Dawn Medley. "For us, it's a start. Once we get the ball rolling, [enrollment] will pick up some momentum."

The state requires public colleges and universities to report their enrollments as of the 11th day of classes. The state Department of Higher Education plans to release its enrollment report for Arkansas' 33 public campuses on Monday, said Steve Floyd, the department's deputy director.

UA's fall enrollment has increased by 20 percent during White's tenure.

Medley said there's still work to do. The campus needs to average a 6.3 percent enrollment increase in each of the next six years to reach White's 2010 goal of 22,500 students.

UA's admissions office added four full-time recruiters in the past 13 months, bringing the total to 10, said Medley, who started work in August 2003. The office became more "customer friendly" for students and spent more on direct-mail advertising, she said.

"We did everything we could to make the U of A work for a student," Medley said.

UA's six-year graduation rate increased to 52.8 percent, up from 48.1 percent. White's goal is to increase the six-year graduation rate to 66 percent by 2010.

The results of the enrollment were preliminary, said Charles Crowson, a UA spokesman. The university is still processing the information, he said.

Officials also released the ethnic figures of UA's freshmen class.

The number of black first year students increased slightly from last year, from 111 to 115.

The 3.6 percentage gain was the lowest among students who identified their ethnicities on UA's applications, according to the enrollment report.

Freshmen who identified themselves as American Indian nearly doubled, from 32 to 62. The number of freshman Asians increased by 4.4 percent and Hispanic freshmen increased nearly 45 percent.

To increase campus diversity, the university introduced a scholarship this fall that honors Silas H. Hunt, the first black student in modern times to attend UA. He entered the UA law school in 1948.

Medley said she was pleased with the overall gains among students who are members of minority groups but said UA needs to focus on attracting more black students.

Overall black enrollment decreased by 1.7 percent last year and only 6.1 percent of the student body was black, compared with a 15.6 percent black population statewide, according to the 2000 Census.

UA is trying to draw more black students from central Arkansas and the Arkansas Delta, Medley said. A new assistant director, Latisha Brunson, will recruit in those areas, Medley said.

Students who declined to identify their ethnicity on UA's application grew 9.5 percent, Medley said.

The university will change its application for the 2005–06 school year and allow a student to report more than one ethnicity, she said.

She hopes students will begin to report their mixed ethnicities and that enrollment for members of minority groups will thus increase, Medley said.

Study Questions for Case Study 5.1

"Enrollment Rates Rise" by Jeff Smith
The Morning News of Northwest Arkansas
September 18, 2004

The article gives information about the increase in enrollments at the University of Arkansas from 1998 to 2004 and comments on the increase in the six-year graduation rate. This set of questions relates to the Smith article and the line graph representing the UA Enrollments.

1. **How many data points are plotted to produce this graph? What do these data points represent?**

2. **Based on the graph, give an estimate of the spring enrollment in 2001. Do you believe this estimate is correct or incorrect? Explain your reasoning.**

3. **What does your answer to #2 say about the choice of a line graph to represent the seven data points? What kind of graph would be a better choice?**

4. **What technique is used in the graph that may exaggerate the increases in enrollment?**

   Comment for thought: When we discuss "enrollment rates," to what exactly are we referring? For example, when we discuss six-year graduation rates, we are referring to the percentage of students who entered the University of Arkansas in a particular year who graduated in the next six years. Relate the definition of graduation rates to the term enrollment rates.

5. **Explain the difference between saying the enrollments rise and saying the enrollment rates rise.**

6. **Is the headline Enrollment Rates Rise descriptive of what the data and article reveal? Explain.**

   Comment for thought: We have thus far discussed two kinds of data: enrollment and enrollment rates. A third kind of data involves both rate and enrollment, but is not usually called an enrollment rate. Rather, the rate of change of enrollment is a measure of how the enrollment changes over time. A standard way of determining this is to compute a rate of change (in this case a percent increase) of the enrollment from one fall to the next fall.

7. **Produce a bar graph of the percent change in the enrollment from 1998–2004.**

8. **Explain how the Enrollment Rates Rise statement could be interpreted correctly. Is it possible for enrollments to increase yet enrollment rates to decrease? Explain your answer.**

9. **What does it mean to say "MORE STUDENTS GRADUATE"? Is the subhead MORE STUDENTS GRADUATE supported by the article? Explain.**

"UA Enrollment up 5% over 2003" by Chris Branam
Arkansas Democrat-Gazette
September 18, 2004

In contrast to the line graph by *The Morning News*, the *Arkansas Democrat-Gazette* showed the same information in a bar graph with a scale of approximately 1 centimeter representing 4,600 students, and the seven bars are reasonably faithful to this scaling.

10. **Is the vertical scale on the second UA enrollment graph correct? Explain your reasoning.**

11. **Which of the two graphs makes the increase in enrollment since 1997 seem larger? Why?**

12. **One of the graphs gives the 1998 enrollment as 14,742 while the other graph gives this as 15,060. Is there any information in either of the two articles that indicates which of these is correct? If so, what is the correct enrollment figure for 1998? Explain.**

13. **Check the mathematics from the headline. Is the UA enrollment up 5% over 2003? Explain.**

14. **Check the mathematics in this statement: "UA's fall enrollment has increased by 20 percent during White's tenure."**

15. **The article states, "The campus needs to average a 6.3 percent enrollment increase in each of the next six years to reach White's 2010 goal of 22,500 students." Is this accurate? Under this rate of change in enrollment, create a table that provides the enrollment for each of the fall semesters from 2005 until 2010.**

Case Study 5.2: Central Plaza Graph

Resource Material: Graphic entitled "Number of Students on Central Plaza on a Typical Week Day."
Origin unknown; brought by a student to class.

Learning Goals: The learning goals of this case study include understanding data presented in unusual graphs and recognizing the patterns of data that are presented in graphs.

This is a study of data presented in an unusual graphical format. The first task is to understand what data are likely being presented. After understanding the graphic, the effectiveness of the presentation is considered.

Article for Case Study 5.2

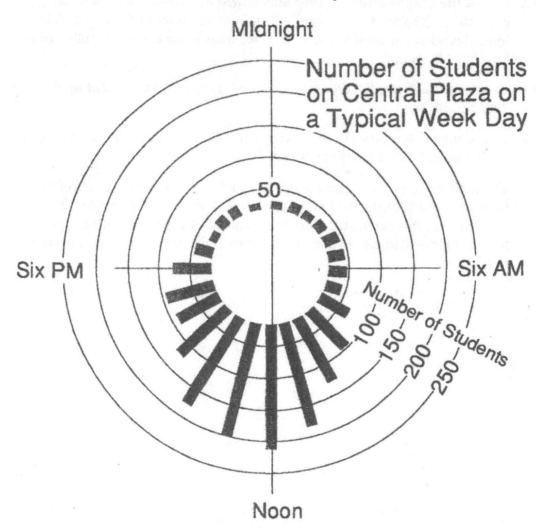

Study Questions for Case Study 5.2

Graphic entitled "Number of Students on Central Plaza on a Typical Week Day"
Origin unknown

1. What data are represented in this graph? Describe how you would interpret this graph. In terms of the data, what is occurring at noon?

2. What trends in the data are immediately evident from looking at the graph? Describe the various trends you see in the data.

3. Would the pattern seen in this graph be different for a Saturday or Sunday? Why?

4. Give approximations of the data that are represented on this graph (each hour) and produce a table AND a standard bar graph of these data.

5. Would a line graph be reasonable to represent these data? Why or why not? Produce a line graph of the data.

6. Discuss the advantages of using the clock graph, the table, the bar graph, and the line graph to represent the data.

7. Think of a setting for which you could generate a similar type of clock graph or line graph (such as the number of people at a local shopping mall or the number of students in the cafeteria at a given time). Try to make your setting somewhat unique, yet something other students would be familiar with and understand. Once you have chosen your setting, create a clock graph or a line graph that represents what you chose. Create a title for your graph similar to the one for the "Number of Students on Central Plaza on a Typical Week Day," and place the title on the back of the graph. Do not share your graph with other students! (see #8)

8. (Class activity) Collect the graphs created by the class and number them. Create a list of the various titles given for these graphs, and then attempt to match graphs with the given titles.

Case Study 5.3: Two Views of a Tax Cut

Resource Material: Graphic containing two graphs representing the 1995 tax cut, *New York Times*, April 7, 1995.

Learning Goals: The learning goals of this case study include reconciling different graphical representations of quantitative information and deriving information from graphical representations.

In 1995, the U.S. Congress enacted a tax cut that was expected to amount to $245 billion over seven years. Different representations of this tax cut were issued by the House Ways and Means Committee and by the Treasury Department. The House Ways and Means Committee produced the Republican Math graph and the Treasury Department produced the Democratic Math graph. In 1995, Republicans had a majority in Congress, and Bill Clinton, a Democrat, was President.

Warm Up Exercises for Case Study 5.3

Suppose that person A has a salary of $200,000 and pays $50,000 in taxes while person B has a salary of $40,000 and pays $5,000 in taxes. A tax cut is enacted that reduces A's taxes by 4% and B's taxes by 5%.

1. What is the total tax reduction for A and B?

2. What percent of the total tax reduction for A and B goes to A?

3. What percent of the total tax reduction for A and B goes to B?

4. Who received the most favorable reduction according to this tax cut? Explain your reasoning.

5. Does this tax cut favor upper-income people like person A or middle-income people like person B? Explain.

6. Would you consider this tax cut fair? Support your reasoning with a quantitative argument.

Article for Case 5.3

Two graphs representing the 1995 tax cut, *New York Times*, **April 7, 1995.**

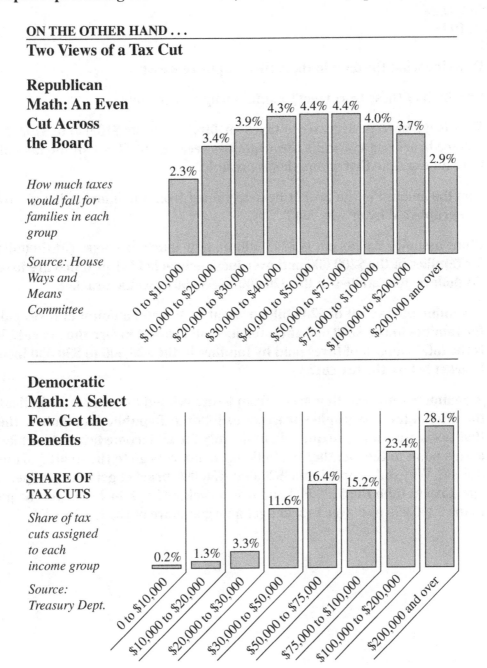

ON THE OTHER HAND ...

Two Views of a Tax Cut

Republican Math: An Even Cut Across the Board

How much taxes would fall for families in each group

Source: House Ways and Means Committee

2.3%   3.4%   3.9%   4.3%   4.4%   4.4%   4.0%   3.7%   2.9%

0 to $10,000
$10,000 to $20,000
$20,000 to $30,000
$30,000 to $40,000
$40,000 to $50,000
$50,000 to $75,000
$75,000 to $100,000
$100,000 to $200,000
$200,000 and over

Democratic Math: A Select Few Get the Benefits

SHARE OF TAX CUTS

Share of tax cuts assigned to each income group

Source: Treasury Dept.

0.2%   1.3%   3.3%   11.6%   16.4%   15.2%   23.4%   28.1%

0 to $10,000
$10,000 to $20,000
$20,000 to $30,000
$30,000 to $50,000
$50,000 to $75,000
$75,000 to $100,000
$100,000 to $200,000
$200,000 and over

Study Questions for Case Study 5.3

"Two Views of a Tax Cut"
New York Times
April 7, 1995

1. Describe what the data in these two graphs represent.

2. Can both of these be correct? Explain why or why not.

3. Why is the percent decrease in taxes for families in the $10,000 to $20,000 income bracket 3.4% while these families receive only 1.3% of the tax cut? Give an example that shows this is possible.

4. Can the amount of the tax cut be determined from the information given in the two graphs? Why or why not?

5. If one assumes the tax cut is $245 billion, how much are taxes cut (in dollars) for families in the $200,000 and over income bracket? How much are taxes cut (in dollars) for families in the $20,000 to $30,000 income bracket.

6. Assuming the tax cut of $245 billion, what is the total amount of taxes paid by the families in the $200,000 and over income bracket before the tax cut? What is the total amount of taxes paid by families in the $20,000 to $30,000 income bracket before the tax cut?

7. A common argument that arises from issues related to tax cuts (as indicated by the titles of the two graphs) proceeds as follows: Republicans allude to the idea that taxes are cut approximately uniformly for all income brackets, while Democrats point to the fact that the bulk of the tax cuts go to the wealthy. How is this so? Why did people in the $20,000–$30,000 bracket get a larger (percentage) tax cut than the $200,000 and over bracket (3.9% to 2.9%) yet the people in the $200,000 and over bracket get a larger share of the tax cuts?

Case Study 5.4: Sexual Betrayal Graph

Resource Material: "The Interaction of Sex of Respondent and Sex of Transgressor on the Acceptance of Sexual Betrayal" graphic. Brought in by a student, from a textbook.

Learning Goals: The learning goals of this case study include understanding data that are represented by a graph.

The data and graph come from a report of a study of 261 college students, 93 males and 168 females.[5] The study and the report are much more extensive than this result, but we focus here on this graph.

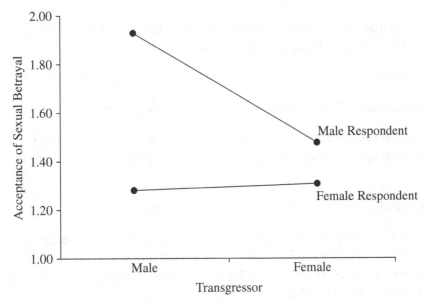

The Interaction of Sex of Respondent and Sex of Transgressor on the Acceptance of Sexual Betrayal

This graph represents how males and females accept sexual betrayal by male and female transgressors. Although not noted on the graph, one assumes that a larger number on the acceptance of sexual betrayal scale (from 1 to 2) indicates a greater degree of acceptance.

5. S. Shirley Feldman, Elizabeth Cauffman, Lene Arnette Jensen, and Jeffery J. Arnett. The unacceptability of betrayal: A study of college students' evaluations of sexual betrayal by a romantic partner and betrayal of a friend's confidence. *Journal of Youth and Adolescence* 29:4 (2000): 498–523.

Study Questions for Case Study 5.4

The Interaction of Sex of Respondent and Sex of Transgressor on the Acceptance of Sexual Betrayal graphic, in "The unacceptability of betrayal: A study of college students' evaluations of sexual betrayal by a romantic partner and betrayal of a friend's confidence" by S. Shirley Feldman, Elizabeth Cauffman, Lene Arnette Jensen, and Jeffery J. Arnett, *Journal of Youth and Adolescence* 29:4 (2000): 498–523.

1. What data are represented in this graph?

2. What do the line segments drawn between the data points (Male Transgressor, 1.27) and (Female Transgressor, 1.28) and between (Male Transgressor, 1.92) and (Female Transgressor, 1.47) represent?

3. From the graph alone, what can you say about the level of acceptance of sexual betrayal by male and female respondents?

4. What possible responses were likely in the survey from which these data come?

5. Construct a table of the data that are used in the graphic.

6. This graphic takes the major portion of a page in a book. The table of data takes far less space. Why would one use the graphic rather than just a table of data?

7. What kind of graphic would better represent the data? Why?

8. Construct two bar graphs of the data from the table with "acceptance" measured on the vertical axis. Use a 0 to 2 scale on one vertical axis and a 0 to 4 scale on the other vertical axis. Which of the bar graphs better describes the outcome of the survey? Why?

Case 5.5: Rising Health Care Costs

Resource Material: "Decade After Health Care Crisis, Soaring Costs Bring New Strains" by Robin Toner and Sheryl Gay Stolberg, *New York Times,* August 11, 2002.

Learning Goals: The learning goals of this case study include distinguishing between the graph of a quantity and the graph of the rate of change of the quantity.

This front-page article had ten graphics that were used to show changes in health care spending, the fraction of the population who are uninsured, and the changes in kinds of insurance coverage. Here we consider the text of the article and some of the accompanying graphics. The changes in spending for health care are shown for each decade over the period 1960 to 2010 (projected), and the changes are given in various components of spending each year for the period 1990 to 2001.

Warm Up Exercises for Case Study 5.5

The following graph illustrates the percent changes in the cost of textbooks during the years 2000–2005. Suppose you spent $800 for textbooks at the beginning of 2000. Give the expected costs at the beginning of the years in a table and produce a graph of these costs. Consider first what kind of graph is best and explain why you chose the type of graph you produced.

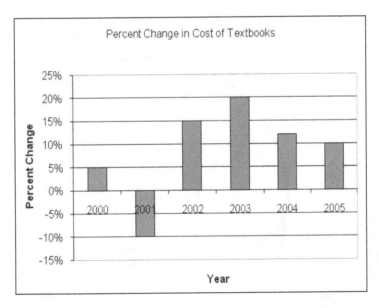

Year	2000	2001	2002	2003	2004	2005	2006
Cost	$800						
Change During Year	5%						xxxx

Article and Graphs for Case Study 5.5

The types of graphs in the article vary. The first (to the right) is a bar graph (one bar for each year) that shows the costs of health care as a percent of the gross domestic product (GDP) over the period 1960 to 2010. A second graph (below) of total costs of health care is presented as a collection of expanding circles, where the costs are measured in constant 1996-dollars.

Five line graphs entitled THE RISE IN SPENDING (below) give annual changes (1990–2001) in percent of total national health care spending, spending for health insurance premiums, costs of prescription drugs, costs of hospital care, and costs for doctors and nurses. Each of these line graphs includes a line graph of the annual inflation rates.

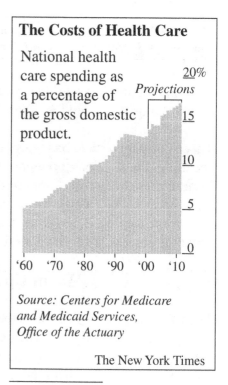

Courtesy of the New York Times Company.

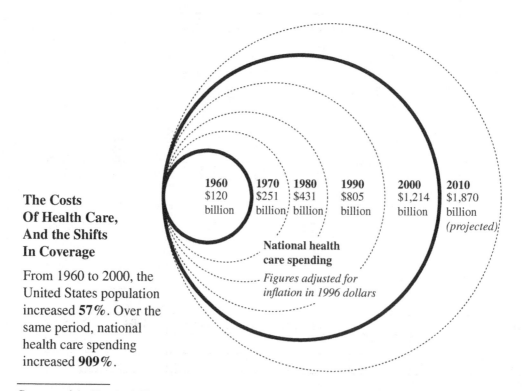

The Costs Of Health Care, And the Shifts In Coverage

From 1960 to 2000, the United States population increased **57%**. Over the same period, national health care spending increased **909%**.

Courtesy of the New York Times Company.

THE RISE IN SPENDING

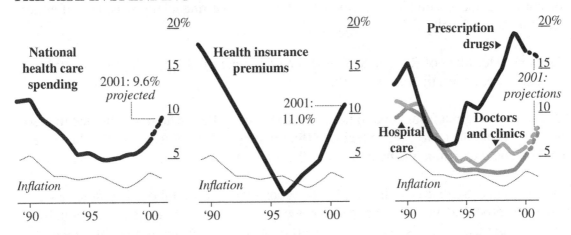

Courtesy of the New York Times Company.

New York Times
August 11, 2002
Decade After Health Care Crisis, Soaring Costs Bring New Strains
By ROBIN TONER and SHERYL GAY STOLBERG

Ten years after a health care crisis threw American politics into turmoil, many experts see another one on the horizon.

The cost of health care, which had stabilized in the mid-1990's with the advent of managed care, is climbing rapidly again, putting new strains on employers, workers and government health programs. In a struggling economy, many employers say they can no longer simply absorb these higher costs and must pass more of them on to employees.

It is not just a problem of rising costs. The troubled economy is expected to cause an increase in the number of Americans without insurance, which stood at 39 million even at the end of the booming 1990's. Families USA, a consumer advocacy group, has estimated that more than two million Americans lost their insurance last year because of layoffs.

If the cost of coverage keeps rising, experts warn, even more Americans will join the ranks of the uninsured because they will be priced out of the market. Many health care analysts, their faith shaken in managed care, see no easy fixes.

Politicians in both parties are beginning to respond, but they are profoundly divided on the issue, a deadlock underscored last month by the Senate's inability to pass a prescription drug benefit for Medicare. As a result, the issue is expected to permeate the fall elections.

Behind the numbers are people like Paul McGonnigal, 36, of Portsmouth, N.H., who lost a six-figure salary and his health benefits when his dot-com startup faltered. "I look at this situation as extraordinarily high risk," said Mr. McGonnigal, now trying to start a consulting business with his wife, Shelley. "We would be financially wiped out if either one of us got seriously ill."

Maryanne McMillan of Richmond, Calif., a city planner disabled by lupus, an autoimmune disease, knows the risk too well.

"My credit has gone from good to hell," said Ms. McMillan, who has struggled to piece together affordable coverage. "I feel like I'm on a high wire act when you're the sickest that you could ever imagine being in your life."

The soaring costs are driven, in part, by the biomedical revolution of the past decade, which has produced an array of expensive new treatments for an aging population, from drugs to fight osteoporosis to high-tech heart pumps. What results is a health care system filled with great promise and inequity, symbolized by wonder drugs that many of the elderly can barely afford.

Dr. Janelle Walhout sees the paradox every day at the community clinic in Seattle where she works. "I've been thinking lately about the mismatch," Dr. Walhout said, "between how very high-tech medicine has become, with all these genetic tests for everything, mixing your medications like fine cocktails, and our patients, who can't afford them, can't understand it, can't get interpreters to explain it and are just not accessing those things."

After the failure of President Bill Clinton's effort to create universal health insurance in 1994, many experts thought that the private sector—health maintenance organizations and other forms of managed care—would deliver cost-efficient, high-quality medical care. But managed care's success in controlling costs proved short-lived, in part because patients and doctors bristled against its hard bargaining and restrictions on care.

Now, many experts agree with Drew Altman, president of the Kaiser Family Foundation, a health research group, who said: "No one has a big new answer on what to do about health care costs. And it's all made worse because health costs are rising in bad economic times."

The strains in the system are increasingly apparent.

Spending on health care rose 6.9 percent in 2000, the largest one-year percentage increase since 1993, federal researchers reported this year. Spending on prescription drugs and hospital care grew particularly fast, largely because of advances in technology and "the retreat from tightly managed care," said Paul B. Ginsburg, president of the Center for Studying Health System Change, a research organization.

Health insurance premiums rose an average of 11 percent last year and are expected to rise an additional 13 percent this year after several years of very modest growth. Premiums for many small businesses will rise even higher, many experts say. The California

Public Employees' Retirement System, or Calpers, reported that its premiums would rise an average of 25 percent next year.

Employers are beginning to pass on those higher costs to their workers, in the form of higher co-payments and deductibles. According to new studies by the Kaiser Foundation, the amounts that employees pay for deductibles in typical health plans rose by more than 30 percent in 2002 from 2001 after little or no growth in recent years.

"Employers are really trying to get back on track in the current economy," said Kate Sullivan, director of health policy at the United States Chamber of Commerce. "But they are seeing their costs just explode."

Denise Mitchell, director of communications for the A.F.L.-C.I.O., said, "In the last six to nine months, health care has become the biggest issue in collective bargaining."

States are struggling with soaring costs in their Medicaid programs, which cover more than 40 million low-income Americans. Governors, squeezed by declining tax revenues, are pleading for more money from Washington. State legislatures are coping by reducing benefits, like dental coverage for adults, capping enrollments and requiring poor people to pay more for their care.

Bruce C. Vladeck, who ran the Medicare and Medicaid programs under Mr. Clinton, describes the situation like this: "The air has gone out of the bubble. We're back into a cycle of cost inflation and an unwillingness, as opposed to the 90's, of employers and to some extent governments to absorb those costs." He added, "They are out of magic bullets."

The picture is not entirely grim. Unemployment is still well below the level it was in the aftermath of the 1991 recession, when middle-class people, frightened that they would lose their health insurance, pushed the issue to the top of the political agenda and helped defeat the first President Bush.

Nor are health care costs rising as rapidly as they did a decade ago—at least so far. Employers' premiums rose an average of 18 percent in 1989, for example, compared with the 13 percent expected this year. At the same time, government actuaries predict that growth in health spending will begin to moderate in a few years.

Moreover, the number of uninsured children has dropped by more than two million since Congress began the Children's Health Insurance Program in 1997. The program helped reduce the percentage of children without health insurance to 10.8 percent from 13.9.

Tommy G. Thompson, the secretary of health and human services, describes that program as "a genuine success story." But even Mr. Thompson says the cost squeeze has left the health care system "stretched and stressed." Many health experts now agree with Joel E. Miller, an analyst at the National Coalition on Health Care, made up of business, labor, consumer and other groups. Mr. Miller has been warning that a "perfect storm" could be brewing.

The Politics–A Clamor for Action But No Consensus

Mindful of the political power of the health care issue in the past, Democrats and Republicans are scrambling to claim it as their own.

But the spectacular failure of the Clinton universal care initiative left politicians unwilling to propose another comprehensive plan. Mr. Clinton's effort to restructure health care, which became a target for an army of interest groups, failed to pass either the House or the Senate, and is widely considered a major reason that his party lost Congress in the 1994 election.

For the rest of the 1990's, both parties pushed step-by-step efforts to expand health coverage and control costs, an approach known as incrementalism. In that time, the number of uninsured grew slowly, dipping only in 1999 and 2000. Most analysts expect the number to rise again when the Census Bureau issues new figures for 2001 in September.

Critics, citing these trends, increasingly say incrementalism has failed.

Among them is the Democratic majority leader, Senator Tom Daschle of South Dakota. Asked if he believed there was a health care crisis, he replied, "Absolutely." Mr. Daschle said he was not wedded to any particular plan but believed that the incremental approach was not enough. "We've got to revisit the issue at a national level," he said.

Mark McClellan, a member of the President's Council of Economic Advisers and senior policy director for health care in the White House, acknowledged, "It is harder for people to get care, the best care, at prices they can afford."

But Dr. McClellan refuses to write off the step-by-step approach. "People who are waiting for the day of a big universal coverage system coming in and taking over all aspects of our health care system are going to be waiting for a long time," he said, "because there's no consensus in this country to do that now."

Indeed, the lack of consensus has bedeviled those trying to tackle health care, and the narrowly divided Congress has consistently deadlocked on the issue in recent years. Lawmakers have spent much of the last three years trying—and failing—to pass legislation giving the 40 million elderly and disabled Americans on Medicare some basic relief for the cost of prescription drugs. The two parties also deadlocked on an economic stimulus package that would have helped laid-off workers trying to keep their health insurance.

The Bush administration has proposed a number of initiatives, including tax credits to help the uninsured buy private coverage, which the White House estimates could result in covering six million to eight million Americans. The administration also backs legislation to limit the amount of damages in medical malpractice verdicts—a step the White House says is necessary to control health costs, including doctors' malpractice premiums.

"What we would like to do is find ways to keep health care affordable while at the same time preserving the U.S. leadership in delivering high quality health care," Dr. McClellan said. "And that is a difficult challenge."

But Mr. Daschle describes the Bush administration as "AWOL" on health care. Democrats in general argue that the administration's tax credits for buying insurance are not nearly enough to reduce the ranks of the uninsured. They tend to support expanding the existing government programs—anathema to many Republicans.

The increase in health costs and the sudden disappearance of the federal budget surplus make it all the harder to undertake a major program.

As Washington fails to act and the public clamors for relief, some political strategists see a widening gap between politicians and voters. Bill McInturff, a Republican pollster, says that in many ways the prescription drug issue is a proxy for the rising concern among voters over the cost of health care generally.

"We keep postponing a larger national debate about a health care system that's far from perfect," Mr. McInturff said, "because it's hard."

Senator Hillary Rodham Clinton of New York acknowledges as much. "People are saying we've got to go back and take a look at this, and I think it needs to be done," Senator Clinton said. But the woman who, as the first lady, produced a 1,342-page plan to restructure the health care system in 1994 now says, "I don't have any road map on how to get there."

The Families–Coverage Climbing Beyond Reach

From factory workers to business executives, Americans are trying to cope with higher health costs. Those who feel it most acutely are people who do not have subsidized group insurance from their employers.

In Nacogdoches, Tex., David and Nicole Alders, who own a small poultry farm and are the parents of seven, have tried to keep their insurance costs down by buying a policy with a $15,000 deductible. Even so, their premiums, now $268 a month, have doubled over the last five years.

"I'm concerned about what will happen five years from now if it doubles again," Mr. Alders said. "A 20 percent annual increase when I'm 40 years old is going to be unaffordable when I'm 60."

In Leesburg, Va., Dale Gardner, who became an independent consultant when he lost his software marketing job, pays $953 a month to cover his family under the federal Cobra program, which allows laid-off workers to keep their benefits for up to 18 months. But as that period comes to a close, Mr. Gardner's wife is considering going back to work—not for the money, he said, but for the insurance.

In Massapequa, N.Y., Trish Patafio, a homemaker, and her husband, Bob, who is self-employed and moves boats for a living, had their family premiums raised to $759 from $559 a month. So the Patafios enrolled their three children in the federally subsidized Children's Health Insurance Program, or CHIP, generally available to families that earn up to twice the federal poverty level, or about $36,000 a year for a family of four.

Even so, Mrs. Patafio said, the couple must pay $514 a month to insure themselves. "We're eating pancakes some nights as it is," she said. "I said to my husband, 'What are we going to do, sell the house so we can pay health insurance?'"

For many, any coverage is out of reach. Dolores Stanfield, 50, a waitress in Columbia, S.C., has not had health insurance for 15 years. Ms. Stanfield suffers from high cholesterol and a spine ailment that requires regular medication and sees a doctor at a government-subsidized clinic, which charges patients based on their incomes. Ms. Stanfield pays $15 to $20 a visit and says she tries to minimize medical tests because "it costs more." Even so, she has $3,000 in outstanding bills at the local emergency room.

About 14 percent of Americans, the vast majority in working families, lack health coverage either because their employers do not offer it or it costs too much. They may earn too much to qualify for government health assistance, or they may be childless adults, and as a result ineligible for many programs.

Ron Pollack, the executive director of Families USA, says that in 42 states a person without children can be penniless and still not qualify for federally subsidized health coverage. Mr. Pollack argues eligibility for public programs should be based on income, not family status.

"This is a political pecking order," Mr. Pollack said, "that somehow children are more popular than parents and parents are more popular than nonparental adults."

That pecking order plays out significantly in the lives of the Everett family of Arlington, Tex.: Corrine Everett, who has multiple sclerosis; her husband, Cliff, who works for a company that offers insurance the family cannot afford; and their two children, ages 12 and 18.

The entire family went without health coverage until two years ago, when the children were enrolled in CHIP. But the parents do not qualify for government help. "He makes too much money for us to get on Medicaid," Mrs. Everett said, "but he doesn't make enough money for us to be able to afford insurance."

Mrs. Everett turned up in the emergency room on a Saturday night three years ago, her arm so numb she could not lift it. It had been bothering her for months, and doctors would later recognize the numbness as a symptom of her multiple sclerosis. But she said she had put off seeking care because she was worried about the cost.

Delaying treatment is typical of the uninsured, according to the Institute of Medicine, an independent research organization. In a recent report, the institute concluded that as many as 18,000 Americans die prematurely each year as a result of not having health coverage. Many wait too long to receive treatment.

In Mrs. Everett's case, she was referred to a neurologist, Dr. Rizwan Shah. Dr. Shah admitted her to the hospital, diagnosed multiple sclerosis and prescribed medicine that costs $1,000 a month. Then the doctor helped enroll her in a charity program sponsored by the drug's maker, which charges her $20 a month for the medication.

If Mrs. Everett's disease flares up, Dr. Shah admits her to the hospital, he said, but always through the emergency room, because emergency rooms are required by law to treat people regardless of whether they can afford to pay. This strategy is more costly for the hospital, he said, "but I have no other choice."

With tears in her eyes, Mrs. Everett told how Dr. Shah charges her only what she can afford. "If he moves on or retires, I'm without a doctor," she said. "I'm not going to find another doctor to do what he does for me."

The Employers–Caught Between Costs And Their Workers

The rising cost of health care has created a complicated blame game among the main players in the system. Patients are blamed for wanting too much care and being heedless of the costs. Managed care companies are blamed for not delivering what they promised: high quality, tightly managed medicine.

Doctors and other health care providers are blamed for "provider pushback"—resisting limits on treatment and asserting their bargaining power for higher reimbursements. Politicians are blamed for undermining these limits by mandating certain benefits and rights of appeal.

In the middle of it all are the employers, who provide health insurance to two-thirds of all Americans under 65. Sears Roebuck & Company provides a glimpse into how businesses are coping.

Sears has always prided itself on offering what Liz Rossman, vice president for benefits, calls "a high level of coverage" under which many treatments and services were available.

When managed care arrived in the early 1990's, promising affordable quality care by focusing on prevention and close management of medical services, the company bought into the idea. It joined an employer group that negotiated with different insurance companies offering a variety of plans.

Health maintenance organizations became the insurance providers of choice at Sears. By the end of the decade, the company was offering about 500 H.M.O. plans, about a half-dozen for each of its locations. Ms. Rossman said 87 percent of the company's 275,000 employees were enrolled.

"Our employees were very, very happy," she said. "It was obviously something that worked."

The company will not provide specific details about what it pays for each plan and what it requires employees to pay, saying it is impossible to generalize.

But about two years ago, Sears officials noticed costs were creeping up again. This year, Ms. Rossman said, some plans asked for as much as a 50 percent increase. The company dropped the most expensive plans, and at the same time began offering what is known as a preferred provider plan under which employees pay a percentage of the cost of their care but have more leeway in choosing their doctors. So far, Ms. Rossman said, the plan is more cost-efficient. Enrollment in traditional H.M.O.'s is now down to 44 percent of the Sears work force.

Ms. Rossman blames insurance companies for the higher costs, saying "most plans didn't have the computer systems or real controls" to manage care and keep costs down.

But Karen Ignagni, president of the American Association of Health Plans, which represents managed care companies, says the rise in health premiums results from a complex tangle of factors, including the increased health care needs of an aging population and advances in medical technology.

Ms. Ignagni resists the argument that managed care has been a failure. She said the system could have succeeded in keeping costs down, had it not been for the rebellion by doctors and other advocates, who fought restrictions on care by lobbying elected officials to require insurers to cover certain tests and procedures, regardless of the cost or whether there is scientific evidence to justify them.

A recent report by Pricewaterhouse Coopers, prepared for Ms. Ignagni's group, says government mandates and regulations are responsible for 15 percent of the recent rise in health care costs.

"There is a political dynamic here," Ms. Ignagni said. She added, "It's all about lobbying."

Not surprisingly, doctors disagree. Dr. Richard Corlin, a former president of the American Medical Association, cited "advancing technology and an aging population," along with the rapid increases in the cost of malpractice insurance, as the primary reasons for the rising cost of care. The A.M.A. also notes that insurance companies are reaping higher profits.

Companies like Sears have been coping with the price increases by asking employees to pay a larger share of the cost of their coverage, in the form of higher deductibles and co-payments. Health care economists call this tactic cost-shifting.

Ms. Sullivan, the health policy expert for the Chamber of Commerce, argues that cost-shifting is not necessarily bad, because it forces patients to take responsibility for what they spend on care. "The patient starts acting like a consumer," Ms. Sullivan said.

Maybe so, says Dr. Ginsburg, of the Center for Studying Health System Change, but sometimes they make bad health care choices in the interest of saving money. "What I worry about," he said, "is cost-sharing becoming a barrier to people getting the care they need."

Even Ms. Sullivan says that if employees are asked to pay too much of the cost, they and their families will simply drop out of employer-sponsored plans, "voluntarily uninsuring themselves."

That is how Adela Velasquez, a part-time housekeeper in San Rafael, Calif., became the only member of her family without health insurance.

Her husband, Francisco Guillen, works for a construction company that offers health benefits to its employees, and pays part of the cost for their families. But the policy is a so-called tiered plan, one that assigns different costs to different family members.

Insuring their four children was a bargain, only $138 per year. But adding Ms. Velasquez to the policy would have cost $3,250 a year.

So Ms. Velasquez roams the pharmacy aisles looking for low-cost over-the-counter remedies when she is sick and visits the doctor only rarely. A year ago, she noticed a painful lump in her groin, but she waited six months before seeking treatment at a public clinic.

For $60 cash, a doctor examined her and diagnosed a hernia, a tear of the groin muscle wall. But the surgery to repair it would cost $7,000, which Ms. Velasquez did not have. Eventually, a surgeon volunteering his service to a private charity performed the operation free.

"We are all healthy, thank God," Ms. Velasquez said, speaking in Spanish through an interpreter. But she added, "For me, there is no insurance."

The States–Struggling to Control Medicaid Spending

For decades, the federal and state governments have been the insurer of last resort for the youngest, oldest and sickest Americans. But the government's safety net programs are under increasing strain, in part because of recent efforts to cover more of the needy.

Nowhere is that strain more apparent than in the states.

Medicaid, the state-administered health insurance program for low-income people, covers one in seven Americans. It pays for 40 percent of the births in the nation and 50 percent of the nursing home care. It is already the second largest item in most state budgets, behind education—and it is growing fast—with state spending on Medicaid up 13 percent last year, according to the National Governors Association.

Even in good times, Medicaid costs would be a burden, and these are not good times. Governors in both parties have been making panicky appeals for more help from the federal government, which pays an average of 57 percent of the Medicaid costs.

But the Bush administration, citing its own budget problems, has rejected their appeals to raise the federal contribution, offering instead to grant states additional flexibility in how they spend Medicaid money. Utah, for example, recently received a federal waiver to reduce benefits to some poor people currently on the rolls so that more low-income people could be offered a bare-bones health package. Mr. Thompson said he had approved 1,950 waivers in his less than two years as secretary of health and human services.

Almost every state is tinkering with Medicaid benefits in an effort to control costs. Some are raising co-payments and requiring the use of generic drugs. Some are freezing payments to doctors and hospitals at current levels.

Fourteen states are reducing benefits, like dental care for adults. Most are looking at how to contain spending for prescription drugs, which rose at an average annual rate of 19.7 percent in the last two years.

Washington State, for example, prided itself on expanding—not contracting—coverage throughout the 1990's. Nearly one in three children in the state are now served by Medicaid and the Children's Health Insurance program.

But the days of expansion are over. With health costs soaring, Dennis Braddock, secretary of the state's Department of Social and Health Services, has proposed a variety of controls on Medicaid that have prompted a wrenching debate.

People on Medicaid would be charged a $5 co-payment if they wanted certain brand-name drugs instead of the generic versions, for example, or a $10 co-payment if they used an emergency room for nonemergency care. The state would also like to freeze enrollment in part of its program when costs exceed budget forecasts, although the poorest would continue to receive health care as an entitlement.

"In general, nobody liked any of the strategies," said Doug Porter, the state's director of Medicaid, after a round of public hearings. "They got very quickly to the human costs."

Already, many doctors in Washington State are declining to accept Medicaid patients because they say the payments are too low. At the Rainier Park Medical Clinic in Seattle, family doctors say that they must sometimes scramble to find specialists who will see their Medicaid patients. Dr. Colin Romero was so grateful to one ophthalmologist who regularly takes his referrals that he said he was thinking of writing a thank-you note.

The Future–Costs May Dictate Change in Approach

On a sticky, rainy afternoon in late July, President Bush took the lectern at a university gymnasium in North Carolina and greeted about 1,700 health care professionals. Mr. Bush had come to propose changing the medical malpractice system. But the backdrop behind him, which read "Strengthening Health Care: Access, Affordability, Accountability," suggested that the White House had a broader message.

"Right now, rising health care costs are undermining the availability of health care, of medical care not only here in North Carolina, but throughout our country," the president said. "And the rising costs were forcing too many people to go without, and that's not right, that is a problem."

The scene was eerily evocative of 1992, when Bill Clinton campaigned by promising every American the right to affordable care. The Bush administration is proposing nothing so bold, but its officials, keenly aware of how this issue hurt the president's father, are careful to voice their concern.

As Mr. Thompson said in a recent interview, "Health care is costing so much, along with prescription drugs, that Congress and this administration have got to address it."

But the question of how to address it is very hard for a society with an insatiable appetite for the newest and most expensive drugs and treatments and a deep resentment of any limits on their health care choices.

That resentment was largely responsible for the death of the Clinton plan, which was brought down, in part, by the fear of government rationing—having elected officials (or their bureaucrats) decide which tests and procedures are covered and which are not. A few years later, the fear of rationing by private corporations—H.M.O.'s and other insurance companies—ignited a patient rebellion against managed care.

Many health policy experts argue that tackling health care costs will require a fundamental cultural shift in the American approach to medicine. They say doctors and patients must begin taking cost into account. They say Americans must limit themselves to treatments that are proven to work and accept the premise that more care does not necessarily mean better care.

"As a society, sooner or later we will have to determine whether there are some benefits that are just too small to justify the cost," said Dr. David Eddy, an independent analyst who advises health care organizations, including the managed care industry. Americans, Dr. Eddy said, "have an enormous tendency to use treatments if we think they work or if we hope they work, even if there is no evidence that they do work."

In the 1990's, for instance, bone marrow transplants were widely used to treat aggressive breast cancer. Then studies showed that the transplants were no better than standard therapy. Hormone replacement therapy, prescribed to millions of American women, has now been discredited as a way to prevent heart disease and stroke.

Dr. Eddy says he believes that a new government agency should be set up to take this kind of scientific literature into account, and then make recommendations about whether new treatments are worth the cost. But while health experts agree there is a critical need for independent evaluations of new technologies, they doubt such an agency will ever come into existence.

"It would be killed by all the lobbying groups," said Uwe Reinhardt, a health economist at Princeton University.

Because Americans cannot agree as a society on what is worth covering—or even whether health care is a basic right—individual families have been left to make their own decisions about what kind of health care they can afford. Those decisions, in the end, are framed by the family's economic situation, Mr. Reinhardt said, creating a system with too little care for families near the bottom and often too much for families near the top.

Mr. Reinhardt is one of many experts who argue that such a system is neither efficient nor just.

No one sees this more clearly than the nation's doctors, among them Dr. Bartolo Barone, a neurosurgeon in Charleston, S.C. When he is in the hospital's emergency room, Dr. Barone said, he sees a system that simply does not make sense.

"We're seeing the consequences of people who are going without their medications for diabetes or high blood pressure," Dr. Barone said. "They are paying the rent, they are buying what the kids need, and they skimp on the medicine. I see the stroke, the heart disease, the expensive consequence of disease that hasn't been properly controlled. If you're worried about controlling costs in health care, this is a foolish way to do it."

But health experts say they are not sure about what comes next.

"We're not going to go back and repeat what we did in the 1990's," said Gail Wilensky, who ran Medicare and Medicaid for the first President Bush. "We're not going to go back to a la carte fee for service of the 1980's."

"We will do something else," Ms. Wilensky said, "and the really interesting question is, 'What will that something else look like?'"

Study Questions for Case Study 5.5

"Decade After Health Care Crisis, Soaring Costs Bring New Strains" by Robin Toner and Sheryl Gay Stolberg
New York Times
August 11, 2002

1. What is the main point of this article (as indicated by the title) that is supported by the graphs? Which graph(s) focus strongly on the main point of the article? Why?

2. At one point in the text of the article, there is a paragraph about a decline in the number of uninsured children.

   a. Determine from the information in this paragraph an estimate of the number of uninsured children.

   b. Is your answer consistent with the estimated number of uninsured people in the U.S. as given in the article? Explain why or why not.

3. Health care costs are projected to increase by approximately 7% per year. Assuming that the $1,870 billion (1996-dollars) for 2010 is correct, create a table for the projected health care costs for the years 2010–2020.

4. If the information about David and Nicole Alders' cost for health insurance premiums is correct and the premiums continue to increase as indicated (double every five years) how much will their monthly insurance premiums be in 20 years? How many times larger will their health insurance premiums be in 20 years compared to information given?

5. Based on the information in the article about Sears Roebuck & Company's health care plans for employees, approximately how many Sears employees were enrolled in HMOs at the time of this article? How many were enrolled in HMOs two years earlier?

6. What factors other than the actual costs of specific health care services are contributing to the increase in dollars spent on health care?

7. Are the factors you identified in #6 taken into account in this article's conclusions that health care costs are soaring? Explain your answer.

8. Using information from the first and second graphs of total health care costs, compute the projected GDP in 2010. What are the units used for this computed GDP?

9. How would you convert the computed projected 2010 GDP from (1996-dollars) to (2000-dollars)? Is there sufficient information in the graphs to convert this to (2010-dollars)?

10. What is the visual effect of the expanding tangent circle graph? When would this effect be useful, and when would it be troublesome?

11. Produce a bar graph that represents the same information as the expanding circle graph. Be sure to give a table of the values you are graphing.

12. Which representation, the bar graph or the circle graph, do you prefer and why?

13. The heading over five of the graphs is THE RISE IN SPENDING. How is the title confusing when looking at the graphs? Is this title appropriate and accurate? Why or why not?

14. Produce a table and a graph of the actual amount of national health care spending over the period 1990 to 2000. Include in the table the percentage change for each year.

15. Assume that the health insurance premiums paid out totaled $5 billion in 1989. Produce a table and a graph of the cost of health insurance premiums over the period 1990–2000.

16. On the same axis as the graph of health insurance premiums spending you created in #15, graph the effect of inflation on the beginning amount of $5 billion and interpret the results.

17. Assume a cost of $200 billion in 1989 instead of a cost of $5 billion. Produce the same graphs as in questions 15 and 16 with this change in the assumed cost in 1989. Interpret the results and compare your graphs from #16 with these new graphs.

Quantitative Reasoning

Section 6

Counting, Probability, Odds, and Risk

In Section 6 we study ways to answer two questions: "How many?" and "What are the chances?" The first question is answered by counting, and direct counting is often impossible or unreasonable. Consequently we study ways to count indirectly.

The second question can be answered several ways—1 out of 10, 10%, 9 to 1 odds, and probability 0.1 are four such examples. We will encounter several terms that are often used but not well understood. Among such terms are "coincidence," "random," and "odds."

We also study the meaning of risk—something that is very much a part of our everyday lives. Our goal is to understand what risk is, investigate several examples of risk, and consider ways risk is reduced.

The content of this section is below.

- Introduction to counting.
- Introduction to probability, odds, and risk.
- Case Study 6.1: Mock Lipitor Ad.
- Case Study 6.2: "Benefits and Risks of Cancer Screening Are Not Always Clear, Experts Say" by Tara Parker-Pope, *New York Times,* October 22, 2009.
- Case Study 6.3: "Why Journalists Can't Add" by Dan Seligman, *Forbes Magazine,* January 21, 2002.
- Case Study 6.4: "The Odds of That" by Lisa Belkin, *New York Times Magazine,* August 11, 2002.

Introduction to Counting

In many cases, counting the number of possibilities of an event or the numbers of various combinations of a set of objects is difficult to do directly. The Counting Principle offers a quick way to determine the number of possibilities of selecting objects from two or more sets.

Counting Principle: If there are k (different) choices for item A and n (different) choices for item B, then there are $k \cdot n$ (different) choices for one of item A and one of item B.

More general: If there are k_1 choices of item A_1, k_2 choices for item A_2, ..., and k_n choices for item A_n, then there are $k_1 \cdot k_2 \cdot \ldots \cdot k_n$ choices for one each of items A_1, A_2, ..., and A_n.

How many ways can we arrange six different letters? For any arrangement, there are six choices for the first letter, five choices for the second letter, four choices for the third letter, and so on. The Counting Principle gives us $6 \cdot 5 \cdot 4 \cdot 3 \cdot 2 \cdot 1 = 720$. Multiplying all the integers from 1 to 6 is often abbreviated 6!, read "6 factorial." More generally $n! = 1 \cdot 2 \cdot 3 \cdot \ldots \cdot (n-1) \cdot n$ and n different objects can be arranged in $n!$ different ways. We call such arrangements **permutations**.

Factorial and Calculator Note: There may be a factorial operation on your calculator. For example, on many TI Graphing Calculators one can look under MATH – PRB. To compute 5! on the calculator put 5 on the home screen and then find the ! operation. Computing with factorials can produce large numbers very fast!

Suppose we have 6 letters, 3 of which are the same. How many different permutations are there? The answer would be 6! if all the letters were distinct. Since there are 3! arrangements (permutations) of the three letters that are the same, we have to divide 6! by 3! to get the number of different arrangements of the 6 letters.

More generally, if we have n objects with r of them identical, then there are $\dfrac{n!}{r!}$ different arrangements of the n objects. Further, if r of the objects are identical and another s of them are identical, then the number of different arrangements is $\dfrac{n!}{r!s!}$. This also generalizes when there are more than two subsets of the objects that are identical.

If we have 8 different objects and we want to choose five of the objects and place them in five spaces, by the counting principle, there are ways $8 \cdot 7 \cdot 6 \cdot 5 \cdot 4$ ways to do that. Note that this is $\dfrac{8!}{(8-5)!}$. We call this the permutations of 8 things taken 5 at a time. We will use the symbol P_5^8 to represent the number of permutations of 8 objects taken 5 at a time. More generally, if we have n different objects and we want to choose r of them ($r \leq n$), there are $\dfrac{n!}{(n-r)!}$ ways to do this. This gives us the following:

The number of permutations of n things taken r at a time is $P_r^n = \dfrac{n!}{(n-r)!}$.

If we are not interested in the arrangement of r objects, that is, order of the objects is not important, we use the phrase, a **combination** of r objects selected from n objects. We use the symbol C_r^n to represent the number of combinations of n objects taken r at a time. Since each combination of r objects has $r!$ different arrangements, it follows that

$$r!C_r^n = P_r^n$$
$$C_r^n = \frac{P_r^n}{r!} = \frac{n!}{r!(n-r)!}$$

Example: Suppose we want to choose a committee of three from a group of 16 people. How many committees are possible? Answer:

$$C_3^{16} = \frac{16!}{3!(16-3)!} = \frac{16 \cdot 15 \cdot 14}{6} = 560.$$

Suppose we want to select a chair, vice-chair, and a member for the three-person committee. How many different committees are there now? (If we interchange the chair and the member, we get a different committee.)

Answer: $P_3^{16} = \dfrac{16!}{(16-3)!} = 3360$

Calculator Note: There may be permutation and combination operations on your calculator. For example, with most TI Graphing Calculators, look under MATH-PRB and find nPr and nCr. To calculate the number of combinations of 16 objects taken 3 at a time on the calculator enter 16nCr3.

Counting the complement: Sometimes it is easier to calculate the number of objects that do not have a certain property than it is to count the number of objects with the property. For example, suppose you want to know how many committees of five people chosen from 15 men and 10 women have at least one woman member. It is quite simple to compute the number of committees that have no women and then subtract that number from the total number of committees. Work that out and then write down how you would do this by counting committees that have one, two, three, four, or five women.

Introduction to Probability, Odds, and Risk

Probability

Assume that one tosses a fair coin, that is, a coin that is just as likely to land H as it is to land T. Since there are two equally likely outcomes of tossing the coin, the (theoretical) probability of H is $\frac{1}{2}$ or 0.5 and the probability of T is also $\frac{1}{2}$ or 0.5. Notice that there are 2 possible outcomes, H and T, and one of those is H, so the probability of H is $\frac{1}{2}$.

If one rolls a (fair) die then each possible roll, a 1, 2, 3, 4, 5, or 6, is equally likely, so the probability of rolling any one of the six values is $\frac{1}{6}$. The number of possible outcomes is six. Suppose one considers the **event** of rolling a number greater than 2. Then the event is rolling a 3, 4, 5, or 6. One can say this by considering the event as the set {3, 4, 5, 6}. Then we say the probability of {3, 4, 5, 6} is the number of elements that are in {3, 4, 5, 6} divided by the total number of outcomes. We use the following notation to summarize this sentence. $P(\{3,4,5,6\}) = \frac{N(\{3,4,5,6\})}{6} = \frac{4}{6} = \frac{2}{3}$. Here $N(\{3, 4, 5, 6\})$ means the number of elements in the set.

In general, if one performs an experiment for which there are n equally likely outcomes, then the **probability of an event E**, which is a subset of the outcomes, is $\frac{N(E)}{n}$. Therefore, in order to compute probabilities, one often has to do a lot of counting.

Example: Consider an experiment of choosing a committee of 5 from 10 women and 15 men. The number of possible committees is $C_5^{25} = \frac{25!}{(25-5)!5!} = \frac{21 \cdot 22 \cdot 23 \cdot 24 \cdot 25}{1 \cdot 2 \cdot 3 \cdot 4 \cdot 5} = 53,130$. Let event M be choosing a committee of all men. The number of ways to do this is $C_5^{15} = \frac{15!}{(15-5)!5!} = \frac{11 \cdot 12 \cdot 13 \cdot 14 \cdot 15}{1 \cdot 2 \cdot 3 \cdot 4 \cdot 5} = 3,003$. So the probability of choosing a committee of all men is $P(M) = \frac{3003}{53130} = .0565$, or about 5.6%. Consequently, the probability of not choosing a committee of all men, which is often denoted as $P(\sim M)$, is $P(\sim M) = 1 - .0565 = .9435$.

Risk

Risk is the exposure to injury or loss. There are two basic measures of risk: the probability of loss and the amount of loss. Consider the following two examples:

a. Death by lightning during the month of July. The probability may be very small, possibly 0.00001 or 1 in 10,000, but the loss, one's life, is great.
b. Loss of $3 used to purchase a lottery ticket. The probability of loss is large, possibly close to 1, but the loss, the $3 purchase price, is small.

Risk is reduced if either the probability of loss is reduced or the amount of the loss is reduced. One can reduce risk by reducing the probability of loss. Examples of reducing risk are installing air bags in an automobile or keeping one's automobile repaired.

Often, risk is assumed in order to have the possibility of a gain. An example of assuming risk for a possible gain would be purchasing a $3 lottery ticket with the hope of gaining millions of dollars.

Insurance is a way of reducing risk by reducing the amount of a loss. For example, consider fire insurance on homes. An insurance company gathers data on the number of homes damaged or destroyed by fire and computes the probabilities of losses by home-owners. Then premiums for policyholders are set so that the total of all premiums will exceed the expected losses of policyholder homes. Each policyholder pays to reduce his/her risk of loss. If a policyholder's home burns down, the insurance company will pay the homeowner an agreed upon amount, thus reducing the homeowner's loss.

Insurance also has another form. For example, an abnormally hot summer may be good for ice cream companies but not so good for poultry growers, while a cool summer would have the opposite effect. Consequently an insurance broker may sell "cool summer" insurance to the ice cream company and "hot summer" insurance to the poultry grower.

Odds

Odds are stated in two ways: Odds in favor of an event and odds against an event.

The odds in favor of an event E compare the probability of E occurring to the probability of E not occurring: $\dfrac{P(E)}{P(\sim E)}$. Here $\sim E$ means "not E," or "E does not happen." Note that $P(\sim E) = 1 - P(E)$. Since this expression of odds is a ratio, an equivalent ratio can be found by multiplying both numbers by the same nonzero number. So, if, for example, $P(E) = 0.2$, then $P(\sim E) = 1 - 0.2 = 0.8$. The odds in favor of E can be stated as $\dfrac{0.2}{0.8}$ or $\dfrac{1}{4}$ or $\dfrac{20}{80}$.

The odds in favor of E can be stated several ways: $P(E)$ to $P(\sim E)$
$P(E) - P(\sim E)$
$P(E){:}P(\sim E)$

The odds against an event E are the reverse or inverse of odds in favor of E. That is, the odds against an event E are: $\dfrac{P(\sim E)}{P(E)}$

For example, if the probability of Team A winning a tournament is 0.25, then the odds in favor of Team A winning are 0.25 to 0.75. These odds are the same as 1 to 3. The odds against Team A winning are 3 to 1 (or 3-1 or 3:1).

Improving the odds

One often hears the expression "increasing (or decreasing) the odds." Some of the exercises and study questions in this section will present instances of this. As an example, consider a basketball team whose star player may miss the upcoming tournament. Without the star player the odds against the team winning the tournament are 10 to 1, and with the star player the odds (in favor of) the team winning the tournament are improved by 25%. The odds in favor of the team winning without the star player are 1 to 10, or, considered as a ratio, $1/10 = 0.1$. Increasing this by 25% gives $1.25 \times 0.1 = 0.125 = 1/8$. Therefore the odds in favor of the team winning if the star player plays are 1 to 8, and the odds against the team winning are 8 to 1.

One can see that increasing the odds that are expressed as 1 to 10 as above by 25% is accomplished by reducing 10 by 25%. If the odds are in a different form, say 5 to 2, then increasing them by 20% is done by changing 5/2 to 2.5 and multiplying by 1.2 getting 3. Hence the increased odds are 3 to 1.

Case Study 6.1: Mock Lipitor Ad

Resource Materials: Mock (reproduction) Ad for drug to reduce the risk of heart attack.

Learning Goals: The learning goals of this case study include using percentages to measure risk and understanding the differences between relative and absolute risk.

Warm Up Exercises for Case Studies 6.1 and 6.2

1. Suppose there is a 10% chance of rain on Monday and a 20% chance of rain on Tuesday. Fill in the blanks:

   a. From Monday to Tuesday, the chance of rain has increased by _____ *percentage points*.

   b. From Monday to Tuesday, the chance of rain has increased by _____ percent.

2. Suppose there is a 40% chance of rain on Wednesday and a 50% chance of rain on Thursday. Fill in the blanks:

   a. From Wednesday to Thursday, the chance of rain has increased by _____ *percentage points*.

   b. From Wednesday to Thursday, the chance of rain has increased by _____ percent.

3. Suppose you are told that one can increase the chance of winning a lottery by 25% if one plays the Washington State lottery instead of the Virginia lottery.

   a. If the chance of winning the Virginia lottery is 1%, what are the chances of winning the Washington State lottery?

   b. If the chance of winning the Virginia lottery is 10%, what are the chances of winning the Washington State lottery?

4. Suppose that there is a 65% chance of having your flight delayed every time you arrive at an airport.

   a. If you arrive at an airport 25 times, about how many flight delays should you expect?

   b. If the chance of a flight delay is reduced by 10% and you arrive at the airport 25 times, about how many flight delays should you expect?

Resource Material for Case Study 6.1

This case study focuses on an advertisement for a drug which claims to reduce the risk of heart attack. A mock version of the ad appears below. To find the original ad, which has appeared in numerous media outlets, try an internet search with the phrase "Lipitor ad".

In patients with multiple risk factors for heart disease,

a New Drug has been found to

reduce the risk of

heart attack

by **36%** *

If you have risk factors such as family history, high blood pressure, age, low HDL ("good" cholesterol) or smoking.

*That means in a large clinical study, 3% of patients taking a sugar pill or placebo had a heart attack compared to 2% of patients taking the New Drug.

Study Questions for Case Study 6.1

1. According to the large print of the ad, how effective is Lipitor in reducing the risk of a heart attack? Explain why the ad emphasizes this quantity by printing it in a large font.

2. Looking into the smaller print, for what type of people has Lipitor been proven to be effective?

3. Looking into the smallest print (content marked by "*"), what were the actual results of the clinical study that provided the original statistic? After reading the smallest print, what is your overall impression of the advertisement?

4. Putting this into perspective: Suppose that this "large clinical study" consisted of 2,000 individuals with multiple risk factors for heart disease. Of these people, 1,000 will be given Lipitor (the experimental group) and 1,000 will be given a placebo (the control group).

   a. Based on the information in the ad, about how many individuals in the control group would you expect to suffer a heart attack?

   b. Based on the information in the ad, about how many individuals in the experimental group would you expect to suffer a heart attack?

   c. How much has Lipitor reduced one's risk of having a heart attack? Report two different measures of this risk reduction and comment on which measure is more meaningful to you. Explain your reasoning.

5. The ad indicates that Lipitor can reduce the risk of heart attack by 36%. Explain how this figure, along with all the other reported figures, could be correct.

Case Study 6.2: Benefits and Risks of Cancer Screening

Resource Material: "Benefits and Risks of Cancer Screening Are Not Always Clear, Experts Say" by Tara Parker-Pope, *New York Times*, October 22, 2009.

Learning Goals: The learning goals of this case study include differentiating between absolute and relative risk, interpreting statements about risk found in the media, and converting statements about risk into alternate forms that may be more useful for individual health decisions.

Article for Case Study 6.2

The New York Times
October 22, 2009
Benefits and Risks of Cancer Screening Are Not Always Clear, Experts Say
By TARA PARKER-POPE

Most people believe that finding cancer early is a certain way to save lives. But the reality of cancer screening is far more complicated.

Studies suggest that some patients are enduring aggressive treatments for cancers that could have gone undetected for a lifetime without hurting them. At the same time, some cancers found through screening and treated in the earliest stages still end up being deadly.

As a result, the chief medical officer for the American Cancer Society now says that the benefits of early detection are often overstated. The cancer society says it will continue to revise its public messages about cancer screening as new information becomes available.

While the limits of cancer screening have long been known in the prevention community, the debate is new and confusing to many patients who have been told repeatedly to undergo screening mammograms or annual blood tests to gauge prostate cancer risk.

"The health professions have played a role in oversimplifying and creating the stage for confusion," said Dr. Barnett S. Kramer, associate director for disease prevention at the National Institutes of Health. "It's important to be clear to the public about what we know and be honest about what we don't know."

Nobody is suggesting that women stop getting mammograms or that men stop discussing prostate cancer screening with their doctor. Instead, the goal is to update public health messages to better reflect the benefits, risks and limits of various forms of cancer screening, Dr. Otis Brawley, chief medical officer of the American Cancer Society, said in an interview.

In a news release issued Wednesday, the cancer society affirmed its current guidelines recommending annual mammography screening for women ages 40 and older, and the group advises men to discuss the risks and benefits of prostate cancer screening with their doctors. But understanding the limitations of screening, the statement said, will help researchers develop better screening tests.

"Cancer is a complicated disease," Dr. Brawley said. "We shouldn't try to fight it with simplistic messages."

Overdiagnosis and overtreatment as a result of cancer screening are a major concern. It is estimated that for every 100 women who are told they have breast cancer, as many as 30 have cancers that are so slow-growing they are unlikely to be life-threatening.

In the case of prostate cancer, for every 100 men with diagnoses, as many as 70 have cancers that if left untreated would never have harmed them. Even for men with aggressive prostate cancer, whether screening improves the odds of survival remains a matter of debate.

"It's not that screening doesn't do any good," said Dr. Laura J. Esserman, a professor of surgery and radiology at the University of California, San Francisco, and co-author of a new analysis of screening risks and benefits in The Journal of the American Medical Association. "But it's not the answer for every kind of cancer, and it's not going to fix all the problems."

"If you get screened, there's a chance you're going to find a cancer that might not be dangerous, and you want to make sure you understand that so you don't get overtreated," Dr. Esserman added.

One goal of the screening community is to communicate cancer statistics better. It is a commonly cited fact that mammography screening for breast cancer lowers a woman's risk of dying from the disease by 20 percent, compared with women who do not get screened. That sounds like a big benefit, but it does not fully communicate the extent to which an individual woman is helped by screening.

Another way to describe the benefits of mammography screening is this: You would have to screen 1,000 women ages 50 and older for 10 years in order to avert one additional death from breast cancer, compared to a similar number of women who are not screened.

For men screening for prostate cancer, the data are less clear. An American study showed no statistical difference in prostate cancer death rates among men who were screened, compared with men who were not. A European study showed that screening reduced the risk of dying from prostate cancer by about 20 percent. But in terms of individual risk, that is not a huge benefit. It means that a man who is not screened has about a 3 percent average risk of dying from prostate cancer. If that man undergoes annual screenings, his risk drops to about 2.4 percent.

Some doctors are worried that restating the public health message about mammography, in particular, will just confuse patients. "We don't want to abandon what we have now or confuse people," said Dr. Marisa Weiss, a breast oncologist and founder of the Web site BreastCancer.org.

Many patients do not understand why screening for cancer might be risky. But for every 1,000 healthy women who undergo annual mammograms, about half will have a stressful false positive within 10 years, and 180 of them will undergo a biopsy.

For men undergoing prostate cancer screening, the downside is even greater. Most prostate cancers occur in older men and are so slow-growing that the patient would die from something else before the cancer became a problem. Yet about 30,000 men do die each year of the disease.

It is impossible to distinguish between harmless prostate cancers and deadly ones. As a result, many of the 200,000 men who receive prostate cancer diagnoses annually are subjected to aggressive treatments that render them incontinent and impotent.

"Patients often are not aware that there are risks associated with cancer screening," said Dr. Therese Bevers, medical director of the cancer prevention center at University of Texas M. D. Anderson Cancer Center. "We need to have more conversations with our patients about that."

Study Questions for Case Study 6.2

"Benefits and Risks of Cancer Screening Are Not Always Clear, Experts Say" by Tara Parker-Pope
New York Times
October 22, 2009

1. According to the article, screening for breast cancer has been found to reduce the risk of dying from this disease. This reduction in risk is described in two different ways.

   a. What are the two descriptions found in the article?

   b. For each description, identify whether the author is describing a *relative change* in risk or an *absolute change* in risk.

   c. Which description gives you the best understanding of the reduction in risk? Why?

2. According to the article, a European study found that screening for prostate cancer has been found to reduce the risk of dying from this disease. This reduction in risk is described in two different ways.

   a. What are the two descriptions found in the article?

   b. For each description, identify whether the author is describing a *relative change* in risk or an *absolute change* in risk.

   c. Which description gives you the best understanding of the reduction in risk? Why?

3. Based on the above statements regarding the effectiveness of cancer screening, is screening more effective for breast cancer or prostate cancer? Explain your reasoning.

4. Fill in the Blanks: Find, and/or compute, the appropriate quantities to make each of the following statements agree with information presented in the article:

   a. Without screening, the risk of dying from prostate cancer is _____ %.

   b. With screening, the risk of dying from prostate cancer is _____ %.

   c. If 1,000 men did not undergo regular screening for prostate cancer, one would expect _____ deaths to occur as a result of this disease.

   d. If 1,000 men did undergo regular screening for prostate cancer, one would expect _____ deaths to occur as a result of this disease.

   e. It can be said that screening for prostate cancer could avert _____ deaths per 1,000 men screened.

5. **Fill in the Blanks:** Find, and/or compute, the appropriate quantities to make each of the following statements agree with information presented in the article:

   a. Without screening, the risk of dying from breast cancer is _____ %.

   b. With screening, the risk of dying from breast cancer is _____ %.

6. A *false positive* is when an initial screening indicates the presence of cancer but after a more careful diagnosis (often more intrusive and costly), it is confirmed that there is no cancer present. Based on the article, what are the chances that a healthy woman receives a false positive for breast cancer? Be sure to provide a clear and complete response.

7. What are some of the drawbacks, concerns, or risks associated with cancer screening?

Case Study 6.3: Why Journalists Can't Add

Resource Material: "Why Journalists Can't Add" by Dan Seligman, *Forbes Magazine*, January 21, 2002.

Learning Goals: The learning goals of this case study include understanding odds, combining and modifying odds, recognizing the misuse of numbers, converting units, and understanding unemployment rates.

The article in this case study notes some quantitative reasoning errors by noted journalists, newspapers and magazines. These errors are made by many non-journalists, and they connect to issues in several case studies in this volume. There are misuses of quantities, misunderstanding of the unemployment rate, confusing bases for percents, confusing large numbers, and flawed interpretations of odds.

Warm Up Exercises for Case Study 6.3

1. How many ways are there to select a three character password in such a way that the first two characters are upper case letters and the third character is a numerical digit? What if the first two characters could be either upper or lower case?

2. Use the factorial (!) notation to represent the number of ways to rearrange 11 different letters.

3. How many ways are there to rearrange the letters in PROBABILITY?

4. One student is to be selected at random from a group of 7 freshmen, 5 sophomores, 8 juniors, and 3 seniors.
   a. What is the probability that the student is either a freshman or a sophomore?
   b. What is the probability that the chosen student is not a freshman?
   c. What are the odds against the chosen student being a freshman?

5. Are the following sets of odds against individual horses winning a race reasonable? Support your answer.
   a. Toughguy 4-1; Smithstreet 3-1; Goaway 2-1; and Winsome 5-2
   b. Tom Lee 3-1; Gravel Road 7-2; Tim Tam 7-1; Crazy Legs 9-1; Run Around 4-1; and Lightning 9-1

6. If the odds against a horse winning a race are 4-1 on a dry track and the odds of the horse winning the race are improved by 20% if the track is muddy, what are the odds against the horse winning the race on a muddy track?

7. The odds against four teams winning a tournament are: Giants 2-1; Flames 3-1; Bears 11-1; and Tigers 2-1.

 a. Find the odds against the winner being the Giants or Tigers.

 b. Find the odds in favor of the winner being the Flames.

Article for Case Study 6.3

Forbes Magazine
January 21, 2002
Why Journalists Can't Add
By DAN SELIGMAN

Liberal arts graduates control the media, which doubtless helps the prose—but generates endless screwups in numbers.

New York Times columnist William Safire is a smart guy, with extraordinary insights about politics, national security or anything else he chooses to write about. But nobody's perfect, and this wordsmith has trouble with numbers, as evidenced in a column last summer that offered the following "early morning line" on 2004 Democratic presidential nominees: Tom Daschle (4-1); Joe Biden (5-1); Richard Gephardt (15-1); John Edwards (9-1); John Kerry (4-1); Pat Leahy (6-1); Joe Lieberman (5-1); Chris Dodd (4-1); Russell Feingold (8-1); and Al Gore (2-1). What Safire doesn't seem to realize is that odds translate into percentage probabilities (e.g., 4-1 means the guy has a 20% chance) and that his probabilities add up to 168%. Alas, mutually exclusive contingencies cannot have probabilities summing up to more than 100%.

After many years of observing media colleagues at work, I would say most of them were standing behind the door when quantitative skills were handed out. They quote T. S. Eliot but are babes in the woods when it comes to correlations or the basic laws of probability. Even when the math is simple, they get bollixed up. *Fortune* recently ran an article on convertible bonds and accompanied it with a table demonstrating that stock prices were far below conversion prices. At eight of the companies, the magazine told us, the stocks were doing so badly that they were selling at more than 100% below the conversion price. How's that? A stock selling below zero? But the scary percentages lent a certain excitement to the copy.

Journalists have trouble distinguishing between the increase or decrease in economic output during a quarter and the annual rate of change. Typical was the *New York Times* editorial following release of last year's third-quarter figures. The editors cited a 0.4% decline in the quarter as a reason for heavy grieving. But that number was, of course, the annual rate, meaning that the decline in the quarter was only 0.1%.

Schools have evidently stopped teaching the difference between "million" and "billion." *The Wall Street Journal* reported Ford's fourth-quarter dividend at $270 billion. The *Times* had some New York City reservoirs holding 63,804 billion gallons. *Business Week* had OPEC cutting production by 1.7 billion barrels a day. Even small numbers are treacherous. A recent howler in FORBES: FedEx packages going down the conveyor at 540 feet per second. (Read "per minute.")

Courtesy of Dan Seligman/Forbes.com.

Another common dereliction is coupling a precise number to an undefined term, as in the *New Yorker* article a year ago worrying about the fact that "Americans spent $27 billion on unproven remedies in 1997." Given the conspicuous absence of agreement about which remedies are "unproven," the statement must be rated meaningless.

High-profile media persons endlessly signal that they themselves don't know what the numbers mean. On Nov. 12 *Washington Post* media critic Howard Kurtz was registering amusement over what he judged to be conflicting reports on unemployment. The *New York Times* had run an article stating "400,000 Americans lost jobs last month," while the *Post*'s own headline said "Unemployment soars by 700,000." Howie neglected to note that the unemployment totals are affected by a lot more than job-losing—by, for example, more people looking for work.

Also recently getting things wrong in the *Post* (and many other papers) was syndicated columnist Ellen Goodman, wrestling with the latest news about mammograms and breast cancer. The news is a study indicating that breast cancer mortality rates do not appear to be reduced by regular mammograms. This is contrary to what women have been told for years and leads Ellen to reflect as follows: "We've been told that picking up cancer on a mammogram, before it's big enough to feel, improves the odds of survival by 30%. Of course I can do the math. Those same figures mean that mammograms make no difference in 70% of the cases." No, she cannot do the math. It is quite consonant with a 30% overall survival gain that mammograms have some benefit in 100% of breast cancer cases.

If only the math majors could write.

Study Questions for Case Study 6.3

"Why Journalists Can't Add" by Dan Seligman
Forbes Magazine
January 21, 2002

1. What does "mutually exclusive contingencies" mean? Give an example of two mutually exclusive contingencies. Give an example of two contingencies that are not mutually exclusive.

2. What is wrong with the list of odds for the potential 2004 Democratic presidential nominees? Support your answer with a calculation and analysis.

3. What are possible meanings of stocks "selling at more than 100% below the conversion price?"

4. Convert 540 feet per second into miles per hour. Is this a reasonable speed for a package conveyer belt?

5. Convert 540 feet per minute into miles per hour. Is this a reasonable speed for a package conveyor belt?

6. Why is the statement that "Americans spend $27 billion on unproven remedies in 1997" cited as a misuse of numbers.

7. Explain why "400,000 Americans lost jobs" and "unemployment soars by 700,000" can both be correct when talking about the employment in the same period.

8. What does it mean to say that regular mammograms improve the odds of survival of breast cancer by 30%. If the odds for survival without regular mammograms are 24-1 and regular mammograms improve these odds by 30%, what are the odds with regular mammograms?

9. What is wrong with Ellen Goodman's interpretation of the statement that regular mammograms improve the odds of surviving breast cancer by 30%?

Case Study 6.4: The Odds of That

Source Material: "The Odds of That" by Lisa Belkin, *New York Times Magazine*, August 11, 2002.

Learning Goals: The learning goals of this case study include understanding the difference between conspiracy and coincidence, understanding the meaning of random events, learning to count indirectly, computing probabilities, and computing odds.

This article centers on the somewhat mysterious deaths over four months of ten or eleven persons all of whom worked in areas associated with bioterror and germ warfare.

The ten biologists who died mysteriously and are clearly identified in the article are:

Benito Que
Don C. Wiley
Vladimir Pasechnik
Robert Schwartz
Set Van Nguyen
Victor Korshunov
Ian Langford
Tanya Holzmayer
David Wynn-Williams
Steven Mostow

The theme of the article is whether or not there is some conspiracy or connection among the deaths or this is merely a coincidence. Consider the following questions.

1. What is coincidence?
2. How does one explain the happening of unusual events?
3. Why are some unusual events, such as being dealt a royal flush, much more noted than more unusual events, such as being dealt the 1, 2, and 3 of clubs, the 6 of spades and the 9 of diamonds?
4. What is the meaning of observed patterns, such as the several occurrences of the number 11, in the September 11, 2001, attacks on the World Trade Center Towers in New York?
5. What does it mean to say that the odds against some event occurring are 1000 to 1?

Warm Up Exercises for Case Study 6.4

1. There are 366 possible birthdays for a person.

   a. How many different possibilities are there for the 5 birthdays of 5 people?

   b. How many possibilities are there for the birthdays of 5 people under the condition that no two of them have the same birthday?

2. What is the probability that a person's birthday is on the first day of a month?

3. What is the probability that 5 people all have different birthdays?

4. What are the odds against at least one common birthday among 5 people?

5. How many possible outcomes are there when flipping a fair coin 5 times? (One possible outcome is HTHTH for the five flips.) What is the probability of the outcome of HHHHH when flipping a coin 5 times? What is the probability of exactly two Hs?

6. How many five-card hands are possible when dealing from a deck of 52 standard playing cards? What is the probability of a royal flush hand? (A royal flush is the A, K, Q, J, 10 of the same suit.) What is the probability of a hand consisting of the 2 and 3 of hearts and the 7, 9, and 10 of spades?

Article for Case Study 6.4

New York Times Magazine, **August 11, 2002**
The Odds of That
By LISA BELKIN

When the Miami Police first found Benito Que, he was slumped on a desolate side street, near the empty spot where he had habitually parked his Ford Explorer. At about the same time, Don C. Wiley mysteriously disappeared. His car, a white rented Mitsubishi Galant, was abandoned on a bridge outside of Memphis, where he had just had a jovial dinner with friends. The following week, Vladimir Pasechnik collapsed in London, apparently of a stroke.

The list would grow to nearly a dozen in the space of four nerve-jangling months. Stabbed in Leesburg, Va. Suffocated in an air-locked lab in Geelong, Australia. Found wedged under a chair, naked from the waist down, in a blood-splattered apartment in Norwich, England. Hit by a car while jogging. Killed in a private plane crash. Shot dead while a pizza delivery man served as a decoy.

What joined these men was their proximity to the world of bioterror and germ warfare. Que, the one who was car-jacked, was a researcher at the University of Miami School of Medicine. Wiley, the most famous, knew as much as anyone about how the immune system responds to attacks from viruses like Ebola. Pasechnik was Russian, and before he defected, he helped the Soviets transform cruise missiles into biological weapons. The chain of deaths—these three men and eight others like them—began last fall, back when emergency teams in moonsuits were scouring the Capitol, when postal workers were dying, when news agencies were on high alert and the entire nation was afraid to open its mail.

In more ordinary times, this cluster of deaths might not have been noticed, but these are not ordinary times. Neighbors report neighbors to the F.B.I.; passengers are escorted off planes because they make other passengers nervous; medical journals debate what to publish, for fear the articles will be read by evil eyes. Now we are spooked and startled by stories like these—all these scientists dying within months of one another, at the precise moment when tiny organisms loom as a gargantuan threat. The stories of these dozen or so deaths started out as a curiosity and were transformed rumor by rumor into the specter of conspiracy as they circulated first on the Internet and then in the mainstream media. What are the odds, after all?

What are the odds, indeed?

For this is not about conspiracy but about coincidence—unexpected connections that are both riveting and rattling. Much religious faith is based on the idea that almost nothing is coincidence; science is an exercise in eliminating the taint of coincidence; police work is often a feint and parry between those trying to prove coincidence and those trying to

---

prove complicity. Without coincidence, there would be few movies worth watching ("Of all the gin joints in all the towns in all the world, she walks into mine"), and literary plots would come grinding to a disappointing halt. (What if Oedipus had not happened to marry his mother? If Javert had not happened to arrive in the town where Valjean was mayor?)

The true meaning of the word is "a surprising concurrence of events, perceived as meaningfully related, with no apparent causal connection." In other words, pure happenstance. Yet by merely noticing a coincidence, we elevate it to something that transcends its definition as pure chance. We are discomforted by the idea of a random universe. Like Mel Gibson's character Graham Hess in M. Night Shyamalan's new movie "Signs," we want to feel that our lives are governed by a grand plan.

The need is especially strong in an age when paranoia runs rampant. "Coincidence feels like a loss of control perhaps," says John Allen Paulos, a professor of mathematics at Temple University and the author of "Innumeracy," the improbable best seller about how Americans don't understand numbers. Finding a reason or a pattern where none actually exists "makes it less frightening," he says, because events get placed in the realm of the logical. "Believing in fate, or even conspiracy, can sometimes be more comforting than facing the fact that sometimes things just happen."

In the past year there has been plenty of conspiracy, of course, but also a lot of things have "just happened." And while our leaders are out there warning us to be vigilant, the statisticians are out there warning that patterns are not always what they seem. We need to be reminded, Paulos and others say, that most of the time patterns that seem stunning to us aren't even there. For instance, although the numbers 9/11 (9 plus 1 plus 1) equal 11, and American Airlines Flight 11 was the first to hit the twin towers, and there were 92 people on board (9 plus 2), and Sept. 11 is the 254th day of the year (2 plus 5 plus 4), and there are 11 letters each in "Afghanistan," "New York City" and "the Pentagon" (and while we're counting, in George W. Bush), and the World Trade towers themselves took the form of the number 11, this seeming numerical message is not actually a pattern that exists but merely a pattern we have found. (After all, the second flight to hit the towers was United Airlines Flight 175, and the one that hit the Pentagon was American Airlines Flight 77, and the one that crashed in a Pennsylvania field was United Flight 93, and the Pentagon is shaped, well, like a pentagon.)

The same goes for the way we think of miraculous intervention. We need to be told that those lucky last-minute stops for an Egg McMuffin at McDonald's or to pick up a watch at the repair shop or to vote in the mayoral primary—stops that saved lives of people who would otherwise have been in the towers when the first plane hit—certainly looked like miracles but could have been predicted by statistics. So, too, can the most breathtaking of happenings—like the sparrow that happened to appear at one memorial service just as a teenage boy, at the lectern eulogizing his mom, said the word "mother." The tiny bird lighted on the boy's head; then he took it in his hand and set it free.

Something like that has to be more than coincidence, we protest. What are the odds? The mathematician will answer that even in the most unbelievable situations, the odds are actually very good. The law of large numbers says that with a large enough denominator—in other words, in a big wide world—stuff will happen, even very weird stuff. "The really unusual day would be one where nothing unusual happens," explains Persi Diaconis, a Stanford statistician who has spent his career collecting and studying examples of coincidence. Given that there are 280 million people in the United States, he says, "280 times a day, a one-in-a-million shot is going to occur."

Throw your best story at him—the one about running into your childhood playmate on a street corner in Azerbaijan or marrying a woman who has a birthmark shaped like a shooting star that is a perfect match for your own or dreaming that your great-aunt Lucy would break her collarbone hours before she actually does—and he will nod politely and answer that such things happen all the time. In fact, he and his colleagues also warn me that although I pulled all examples in the prior sentence from thin air, I will probably get letters from readers saying one of those things actually happened to them.

And what of the deaths of nearly a dozen scientists? Is it really possible that they all just happened to die, most in such peculiar, jarring ways, within so short a time? "We can never say for a fact that something isn't a conspiracy," says Bradley Efron, a professor of statistics at Stanford. "We can just point out the odds that it isn't."

I first found myself wondering about coincidence last spring when I read a small news item out of the tiny Finnish town of Raahe, which is 370 miles north of Helsinki. On the morning of March 5, two elderly twin brothers were riding their bicycles, as was their habit, completing their separate errands. At 9:30, one brother was struck by a truck along coastal Highway 8 and killed instantly. About two hours later and one mile down the same highway, the other brother was struck by a second truck and killed.

"It was hard to believe this could happen just by chance," says Marko Salo, the senior constable who investigated both deaths for the Raahe Police Department. Instead, the department looked for a cause, thinking initially that the second death was really a suicide.

"Almost all Raahe thought he did it knowing that his brother was dead," Salo says of the second brother's death. "They thought he tried on purpose. That would have explained things." But the investigation showed that the older brother was off cheerfully getting his hair cut just before his own death.

The family could not immediately accept that this was random coincidence, either. "It was their destiny," offers their nephew, who spoke with me on behalf of the family. It is his opinion that his uncles shared a psychic bond throughout their lives. When one brother became ill, the other one fell ill shortly thereafter. When one reached to scratch his nose, the other would often do the same. Several years ago, one brother was hit and injured by a car (also while biking), and the other one developed pain in the same leg.

The men's sister had still another theory entirely. "She worried that it was a plot to kill both of them," the nephew says, describing his aunt's concerns that terrorists might have made their way to Raahe. "She was angry. She wanted to blame someone. So she said the chances of this happening by accident are impossible."

Not true, the statisticians say. But before we can see the likelihood for what it is, we have to eliminate the distracting details. We are far too taken, Efron says, with superfluous facts and findings that have no bearing on the statistics of coincidence. After our initial surprise, Efron says that the real yardstick for measuring probability is "How surprised should we be?" How surprising is it, to use this example, that two 70-year-old men in the same town should die within two hours of each other? Certainly not common, but not unimaginable. But the fact that they were brothers would seem to make the odds more astronomical. This, however, is a superfluous fact. What is significant in their case is that two older men were riding bicycles along a busy highway in a snowstorm, which greatly increases the probability that they would be hit by trucks.

Statisticians like Efron emphasize that when something striking happens, it only incidentally happens to us. When the numbers are large enough, and the distracting details are removed, the chance of anything is fairly high. Imagine a meadow, he says, and then imagine placing your finger on a blade of grass. The chance of choosing exactly that blade of grass would be one in a million or even higher, but because it is a certainty that you will choose a blade of grass, the odds of one particular one being chosen are no more or less than the one to either side.

Robert J. Tibshirani, a statistician at Stanford University who proved that it was probably not coincidence that accident rates increase when people simultaneously drive and talk on a cellphone, leading some states to ban the practice, uses the example of a hand of poker. "The chance of getting a royal flush is very low," he says, "and if you were to get a royal flush, you would be surprised. But the chance of any hand in poker is low. You just don't notice when you get all the others; you notice when you get the royal flush."

When these professors talk, they do so slowly, aware that what they are saying is deeply counterintuitive. No sooner have they finished explaining that the world is huge and that any number of unlikely things are likely to happen than they shift gears and explain that the world is also quite small, which explains an entire other type of coincidence. One relatively simple example of this is "the birthday problem." There are as many as 366 days in a year (accounting for leap years), and so you would have to assemble 367 people in a room to absolutely guarantee that two of them have the same birthday. But how many people would you need in that room to guarantee a 50 percent chance of at least one birthday match?

Intuitively, you assume that the answer should be a relatively large number. And in fact, most people's first guess is 183, half of 366. But the actual answer is 23. In Paulos's book, he explains the math this way: "[T]he number of ways in which five dates can be chosen (allowing for repetitions) is $(365 \times 365 \times 365 \times 365 \times 365)$. Of all these 365^5 ways, however, only $(365 \times 364 \times 363 \times 362 \times 361)$ are such that no two of the dates are the same; any of the 365 days can be chosen first, any of the remaining 364 can be chosen second and so on. Thus, by dividing this latter product ($365 \times 364 \times 363 \times 362 \times 361$) by 365^5, we get the probability that five persons chosen at random will have no birthday in common. Now, if we subtract this probability from 1 (or from 100 percent if we're dealing with percentages), we get the complementary probability that at least two of the five people do have a birthday in common. A similar calculation using 23 rather than 5 yields 1/2, or 50 percent, as the probability that at least 2 of 23 people will have a common birthday."

Got that?

Using similar math, you can calculate that if you want even odds of finding two people born within one day of each other, you only need 14 people, and if you are looking for birthdays a week apart, the magic number is seven. (Incidentally, if you are looking for an even chance that someone in the room will have your exact birthday, you will need 253 people.) And yet despite numbers like these, we are constantly surprised when we meet a stranger with whom we share a birth date or a hometown or a middle name. We are amazed by the overlap—and we conveniently ignore the countless things we do not have in common.

Which brings us to the death of Benito Que, who was not, despite reports to the contrary, actually a microbiologist. He was a researcher in a lab at the University of Miami Sylvester Cancer Center, where he was testing various agents as potential cancer drugs. He never worked with anthrax or any infectious disease, according to Dr. Bach Ardalan, a professor of medicine at the University of Miami and Que's boss for the past three years. "There is no truth to the talk that Benito was doing anything related to microbiology," Ardalan says. "He certainly wasn't doing any sensitive kind of work that anyone would want to hurt him for."

But those facts got lost amid the confusion—and the prevalence of very distracting details—in the days after he died. So did the fact that he had hypertension. On the afternoon of Monday, Nov. 19, Que attended a late-afternoon lab meeting, and as it ended, he mentioned that he hadn't been feeling well. A nurse took Que's blood pressure, which was 190/110. "I wanted to admit him" to the hospital, Ardalan says, but Que insisted on going home.

Que had the habit of parking his car on Northwest 10th Avenue, a side street that Ardalan describes as being "beyond the area considered to be safe." His spot that day was in front of a house where a young boy was playing outside. Four youths approached Que as he neared his car, the boy later told the police, and there might have been some baseball bats involved. When the police arrived, they found Que unconscious. His briefcase was at his side, but his wallet was gone. His car was eventually found abandoned several miles from the scene. He was taken to the hospital, the same one at which he worked, where he spent more than a week in a coma before dying without ever regaining consciousness.

The mystery, limited to small items in local Florida papers at first, was "What killed Benito Que?" Could it have been the mugging? A CAT scan showed no signs of bony fracture. In fact, there were no scrapes or bruises or other physical signs of assault. Perhaps he died of a stroke? His brain scan did show a "huge intracranial bleed," Ardalan says, which would have explained his earlier headache, and his high blood pressure would have made a stroke likely.

In other words, this man just happened to be mugged when he was a stroke waiting to be triggered. That is a jarring coincidence, to be sure. But it is not one that the world was likely to have noticed if Don Wiley had not up and disappeared.

Don C. Wiley was a microbiologist. He did some work with anthrax, and a lot of work with H.I.V., and he was also quite familiar with Ebola, smallpox, herpes and influenza. At 57, he was the father of four children and a professor of biochemistry and biophysics in the department of molecular and cellular biology at Harvard.

On Nov. 15, four days before the attack on Benito Que, Wiley was in Memphis to visit his father and to attend the annual meeting of the scientific advisory board of St. Jude's Research Hospital, of which he was a member. At midnight, he was seen leaving a banquet at the Peabody Hotel in downtown Memphis. Friends and colleagues say he had a little to drink but did not appear impaired, and they remember him as being in a fine mood, looking forward to seeing his wife and children, who were about to join him for a short vacation.

Wiley's father lives in a Memphis suburb, and that is where Wiley should have been headed after the banquet. Instead, his car was found facing in the opposite direction on the Hernando DeSoto Bridge, which spans the Mississippi River at the border of Tennessee and Arkansas. When the police found the car at 4 a.m., it was unlocked, the keys were in the ignition and the gas tank was full. There was a scrape of yellow paint on the driver's side, which appeared to come from a construction sign on the bridge, and a right hubcap was missing on the passenger side, where the wheel rims were also scraped. There was no sign, however, of Don Wiley.

The police trawled the muddy Mississippi, but they didn't really expect to find him. Currents run fast at that part of the river, and a body would be quickly swept away. At the start of the search, they thought he might have committed suicide; others had jumped from the DeSoto Bridge over the years. Detectives searched Wiley's financial records, his family relationships, his scientific research—anything for a hint that the man might have had cause to take his own life.

Finding nothing, the investigation turned medical. Wiley, they learned, had a seizure disorder that he had hidden from all but family and close friends. He had a history of two or three major episodes a year, his wife told investigators, and the condition was made worse when he was under stress or the influence of alcohol. Had Wiley, who could well have been tired, disoriented by bridge construction and under the influence of a few drinks, had a seizure that sent him over the side of the bridge?

That was the theory the police spoke of in public, but they were also considering something else. The week that Wiley disappeared coincided with the peak of anthrax fear throughout the country. Tainted letters appeared the month before at the Senate and the House of Representatives. Two weeks earlier, a New York City hospital worker died of inhaled anthrax. Memphis was not untouched by the scare; a federal judge and two area congressmen each received hoax letters. Could it be mere chance that this particular scientist, who had profound knowledge of these microbes, had disappeared at this time?

"The circumstances were peculiar," says George Bolds, a spokesman for the Memphis bureau of the F.B.I., which was called in to assist. "There were questions that had to be asked. Could he have been kidnapped because his scientific abilities would have made him capable of creating anthrax? Or maybe he'd had some involvement in the mailing of the anthrax, and he'd disappeared to cover his tracks? Did his co-conspirators grab him and kill him?

"We were in new territory," Bolds continued. "Just because something is conceivable doesn't mean it's actually happened, but at the same time, just because it's never happened before doesn't mean it can't happen. People's ideas of what is possible definitely changed on Sept. 11. People feel less secure and less safe. I'm not sure that they're at greater risk than they were before. Maybe they're just more aware of the risk they are actually at."

As a species, we appear to be biologically programmed to see patterns and conspiracies, and this tendency increases when we sense that we're in danger. "We are hard-wired to overreact to coincidences," says Persi Diaconis. "It goes back to primitive man. You look in the bush, it looks like stripes, you'd better get out of there before you determine the odds that you're looking at a tiger. The cost of being flattened by the tiger is high. Right now, people are noticing any kind of odd behavior and being nervous about it."

Adds John Allen Paulos: "Human beings are pattern-seeking animals. It might just be part of our biology that conspires to make coincidences more meaningful than they really are. Look at the natural world of rocks and plants and rivers: it doesn't offer much evidence for superfluous coincidences, but primitive man had to be alert to all anomalies and respond to them as if they were real."

For decades, all academic talk of coincidence has been in the context of the mathematical. New work by scientists like Joshua B. Tenenbaum, an assistant professor in the department of brain and cognitive sciences at M.I.T., is bringing coincidence into the realm of human cognition. Finding connections is not only the way we react to the extraordinary, Tenenbaum postulates, but also the way we make sense of our ordinary world. "Coincidences are a window into how we learn about things," he says. "They show us how minds derive richly textured knowledge from limited situations."

To put it another way, our reaction to coincidence shows how our brains fill in the factual blanks. In an optical illusion, he explains, our brain fills the gaps, and although people take it for granted that seeing is believing, optical illusions prove that's not true.

"Illusions also prove that our brain is capable of imposing structure on the world," he says. "One of the things our brain is designed to do is infer the causal structure of the world from limited information."

If not for this ability, he says, a child could not learn to speak. A child sees a conspiracy, he says, in that others around him are obviously communicating and it is up to the child to decode the method. But these same mechanisms can misfire, he warns. They were well suited to a time of cavemen and tigers and can be overloaded in our highly complex world. "It's why we have the urge to work everything into one big grand scheme," he says. "We do like to weave things together."

"But have we evolved into fundamentally rational or fundamentally irrational creatures? That is one of the central questions."

We pride ourselves on being independent and original, and yet our reactions to nearly everything can be plotted along a predictable spectrum. When the grid is coincidences, one end of the scale is for those who believe that these are entertaining events with no meaning; at the other end are those who believe that coincidence is never an accident.

The view of coincidence as fate has lately become something of a minitrend in the New Age section of bookstores. Among the more popular authors is SQuire Rushnell (who, in the interest of marketing, spells his first name with a capital Q). Rushnell spent 20 years producing such television programs as "Good Morning America" and "Schoolhouse Rock." His fascination with coincidence began when he learned that both John Adams and Thomas Jefferson died on the same July 4, 50 years after the ratification of the Declaration of Independence.

"That stuck in my craw," Rushnell says, "and I couldn't stop wondering what that means." And so Rushnell wrote "When God Winks: How the Power of Coincidence Guides Your Life." The book was published by a small press shortly before Sept. 11 and sold well without much publicity. It will be rereleased with great fanfare by Simon & Schuster next month. Its message, Rushnell says, is that "coincidences are signposts along your universal pathway. They are hints that you are going in the right direction or that you should change course. It's like your grandmother sitting across the Thanksgiving table from you and giving you a wink. What does that wink mean? 'I'm here, I love you, stay the course.'"

During my interview with Rushnell, I told him the following story: On a frigid December night many years ago, a friend dragged me out of my warm apartment, where I planned to spend the evening in my bathrobe nursing a cold. I had to come with her to the movies, she said, because she had made plans with a pal from her office, and he was bringing a friend for me to meet. Translation: I was expected to show up for a last-minute blind date. For some reason, I agreed to go, knocking back a decongestant as I left home. We arrived at the theater to find that the friend who was supposed to be my "date" had canceled, but not to worry, another friend had been corralled as a replacement. The replacement and I both fell asleep in the movie (I was sedated by cold medicine; he was a medical resident who had been awake for 36 hours), but four months later we were engaged, and we have been married for nearly 15 years.

Rushnell was enthralled by this tale, particularly by the mystical force that seemed to have nudged me out the door when I really wanted to stay home and watch "The Golden Girls." I know that those on the other end of the spectrum—the scientists and mathematicians—would have offered several overlapping explanations of why it was unremarkable.

There are, of course, the laws of big numbers and small numbers—the fact that the world is simultaneously so large that anything can happen and so small that weird things seem to happen all the time. Add to that the work of the late Amos Tversky, a giant in the field of coincidence theory, who once described his role in this world as "debugging human intuition." Among other things, Tversky disproved the "hot hand" theory of basketball, the belief that a player who has made his last few baskets will more likely than not make his next. After examining thousands of shots by the Philadelphia 76ers, he proved that the odds of a successful shot cannot be predicted by the shots that came before.

Tversky similarly proved that arthritis sufferers cannot actually predict the weather and are not in more pain when there's a storm brewing, a belief that began with the ancient Greeks. He followed 18 patients for 15 months, keeping detailed records of their reports of pain and joint swelling and matching them with constantly updated weather reports. There was no pattern, he concluded, though he also conceded that his data would not change many people's beliefs.

We believe in such things as hot hands and arthritic forecasting and predestined blind dates because we notice only the winning streaks, only the chance meetings that lead to romance, only the days that Grandma's hands ache before it rains. "We forget all the times that nothing happens," says Ruma Falk, a professor emeritus of psychology at the Hebrew University in Jerusalem, who studied years ago with Tversky. "Dreams are another example," Falk says. "We dream a lot. Every night and every morning. But it sometimes happens that the next day something reminds you of that dream. Then you think it was a premonition."

Falk's work is focused on the question of why we are so entranced by coincidence in the first place. Her research itself began with a coincidence. She was on sabbatical in New York from her native Israel, and on the night before Rosh Hashana she happened to meet a friend from Jerusalem on a Manhattan street corner. She and the friend stood on that corner and marveled at the coincidence. What is the probability of this happening? she remembers wondering. What did this mean?

"How stupid we were," Falk says now, "to be so surprised. We related to all the details that had converged to create that moment. But the real question was what was the probability that at some time in some place I would meet one of my circle of friends? And when I told this story to others at work, they encoded the events as two Israelis meeting in New York, something that happens all the time."

Why was her experience so resonant for her, Falk asked herself, but not for those around her? One of the many experiments she has conducted since then proceeded as follows: she visited several large university classes, with a total of 200 students, and asked each student to write his or her birth date on a card. She then quietly sorted the cards and found the handful of birthdays that students had in common. Falk wrote those dates on the blackboard. April 10, for instance, Nov. 8, Dec. 16. She then handed out a second card and asked all the students to use a scale to rate how surprised they were by these coincidences.

The cards were numbered, so Falk could determine which answers came from respondents who found their own birth date written on the board. Those in that subgroup were consistently more surprised by the coincidence than the rest of the students. "It shows the stupid power of personal involvement," Falk says.

The more personal the event, the more meaning we give it, which is why I am quite taken with my story of meeting my husband (because it is a pivotal moment in my life), and why SQuire Rushnell is also taken with it (because it fits into the theme of his book), but also why Falk is not impressed at all. She likes her own story of the chance meeting on a corner better than my story, while I think her story is a yawn.

The fact that personal attachment adds significance to an event is the reason we tend to react so strongly to the coincidences surrounding Sept. 11. In a deep and lasting way, that tragedy feels as if it happened to us all.

Falk's findings also shed light on the countless times that pockets of the general public find themselves at odds with authorities and statisticians. Her results might explain, for instance, why lupus patients are certain their breast implants are the reason for their illness, despite the fact that epidemiologists conclude there is no link, or why parents of autistic children are resolute in their belief that childhood immunizations or environmental toxins or a host of other suspected pathogens are the cause, even though experts are skeptical. They might also explain the outrage of all the patients who are certain they live in a cancer cluster, but who have been told otherwise by researchers.

Let's be clear: this does not mean that conspiracies do not sometimes exist or that the environment never causes clusters of death. And just as statistics are often used to show us that we should not be surprised, they can also prove what we suspect, that something is wrong out there.

"The fact that so many suspected cancer clusters have turned out to be statistically insupportable does not mean the energy we spent looking for them has been wasted," says Dr. James M. Robins, a professor of epidemiology and biostatistics at Harvard and an expert on cancer clusters. "You're never going to find the real ones if you don't look at all the ones that don't turn out to be real ones."

Most often, though, coincidence is a sort of Rorschach test. We look into it and find what we already believe. "It's like an archer shooting an arrow and then drawing a circle around it," Falk says. "We give it meaning because it does mean something—to us."

Vladimir Pasechnik was 64 when he died. His early career was spent in the Soviet Union working at Biopreparat, the site of that country's biological weapons program. He defected in 1989 and spilled what he knew to the British, revealing for the first time the immense scale of Soviet work with anthrax, plague, tularemia and smallpox.

For the next 10 years, he worked at the Center for Applied Microbiology and Research, part of Britain's Department of Health. Two years ago, he left to form Regma Biotechnologies, whose goal was to develop treatment for tuberculosis and other infectious disease. In the weeks before he died, Pasechnik had reportedly consulted with authorities about the growing anthrax scare. Despite all these intriguing details, there is nothing to suggest that his death was caused by anything other than a stroke.

Robert Schwartz's death, while far more dramatic and bizarre, also appears to have nothing to do with the fact that he was an expert on DNA sequencing and analysis. On Dec. 10 he was found dead on the kitchen floor of his isolated log-and-fieldstone farmhouse near Leesburg, Va., where he had lived alone since losing his wife to cancer four years ago and his children to college. Schwartz had been stabbed to death with a two-foot-long sword, and his killer had carved an X on the back of his neck.

Three friends of Schwartz's college-age daughter were soon arrested for what the prosecutor called a "planned assassination"; two of the trials for first-degree murder are scheduled for this month. A few weeks later, police arrested the daughter as well. One suspect has a history of mental illness, and their written statements to police talk of devil worship and revenge. There is no talk, however, of microbiology.

On the same day that Schwartz died, Set Van Nguyen, 44, was found dead in an air-locked storage chamber at the Australian Commonwealth's Scientific and Industrial Research Organization's animal diseases facility in Geelong. A months-long internal investigation concluded that a string of equipment failures had allowed nitrogen to build up in the room, causing Nguyen to suffocate. Although the center itself dealt with microbes like mousepox, which is similar to smallpox, Nguyen himself did not. "Nguyen was in no way involved in research into mousepox," says Stephen Prowse, who was the acting director of the Australian lab during the investigation. "He was a valued member of the laboratory's technical support staff and not a research scientist."

Word of all these deaths (though not the specific details) found its way to Ian Gurney, a British writer. Gurney is the author of "The Cassandra Prophecy: Armageddon Approaches," a book that uses clues from the Bible to calculate that Judgment Day will occur in or about the year 2023. He is currently researching his second book, which is in part about the threat of nuclear and biological weapons, and after Sept. 11 he entered a news alert request into Yahoo, asking to be notified whenever there was news with the key word "microbiologist."

First Que, then Wiley, then Pasechnik, Schwartz and Nguyen popped up on Gurney's computer. "I'm not a conspiracy theorist," says the man who has predicted the end of the world, "but it certainly did look suspicious." Gurney compiled what he had learned from these scattered accounts into an article that he sent to a number of Web sites, including Rense.com, which tracks U.F.O. sightings worldwide. "Over the past few weeks," Gurney wrote, "several world-acclaimed scientific researchers specializing in infectious diseases and biological agents such as anthrax, as well as DNA sequencing, have been found dead or have gone missing."

The article went on to call Benito Que, the cancer lab technician, "a cell biologist working on infectious diseases like H.I.V.," and said that he had been attacked by four men with a baseball bat but did not mention that he suffered from high blood pressure. It then described the disappearance of Wiley without mentioning his seizure disorder and the death of Pasechnik without saying that he had suffered a stroke. It gave the grisly details of Schwartz's murder, but said nothing of the arrests of his daughter's friends. Nguyen, in turn, was described as "a skilled microbiologist," and it was noted that he shared a last name with Kathy Nguyen, the 61-year-old hospital worker who just happened to be the one New Yorker to die of anthrax.

Of course, there have always been rumors based on skewed historical fact. Recall, for example, the list of coincidences that supposedly linked the deaths of Presidents Lincoln and Kennedy. It goes, in part, like this: The two men were elected 100 years apart; their assassins were born 100 years apart (in fact, 101 years apart); they were both succeeded by men named Johnson; and the two Johnsons were born 100 years apart. Their names each contain seven letters; their successors' names each contain 13 letters; and their assassins' names each contain 15 letters. Lincoln was shot in a theater and his assassin ran to a warehouse, while Kennedy was shot from a warehouse and his assassin ran to a theater. Lincoln, or so the story goes, had a secretary named Kennedy who warned him not to go to the theater the night he was killed (for the record, Lincoln's White House secretaries were named John Nicolay and John Hay, and Lincoln regularly rejected warnings not to attend public events out of fear for his safety, including his own inauguration); Kennedy, in turn, had a secretary named Lincoln (true, Evelyn Lincoln) who warned him not to go to Dallas (he, too, was regularly warned not to go places, including San Antonio the day before his trip to Dallas).

I first read about these connections five years after the Kennedy assassination, when I was 8, which says something about how conspiracy theory speaks to the child in all of us. But it also says something about the technology of the time. The numerological coincidences from the World Trade Center that I mentioned at the start of this article made their way onto my computer screen by Sept. 15, from a friend of a friend of a friend of an acquaintance, ad infinitum and ad nauseam.

Professor Robins of Harvard points out that "the Web has changed the scale of these things." Had there been a string of dead scientists back in 1992 rather than 2002, he says, it is possible that no one would have ever known. "Back then, you would not have had the technical ability to gather all these bits and pieces of information, while today you'd be able to pull it off. It's well known that if you take a lot of random noise, you can find chance patterns in it, and the Net makes it easier to collect random noise."

The Gurney article traveled from one Web site to the next and caught the attention of Paul Sieveking, a co-editor of Fortean Times, a magazine that describes itself as "the Journal of Strange Phenomena."

"People send me stuff all the time," Sieveking says. "This was really interesting." Wearing his second hat as a columnist for the The Sunday Telegraph in London, he wrote a column on the subject for that paper titled "Strange but True—The Deadly Curse of the Bioresearchers." His version began with the link between the two Nguyens and concluded, "It is possible that nothing connects this string of events, but . . . it offers ample fodder for the conspiracy theorist or thriller writer."

Commenting on the story months later, Sieveking says: "It's probably just a random clumping, but it just happens to look significant. We're all natural storytellers, and conspiracy theorists are just frustrated novelists. We like to make up a good story out of random facts."

Over the months, Gurney added names to his list and continued to send it to virtual and actual publications around the U.S. Mainstream newspapers started taking up the story, including an alternative weekly in Memphis, where interest in the Wiley case was particularly strong, and most recently The Toronto Globe and Mail. The tally of "microbiologists" is now at 11, give or take, depending on the story you read. In addition to the men already discussed, the names that appear most often are these: Victor Korshunov, a Russian expert in intestinal bacteria, who was bashed over the head near his home in Moscow; Ian Langford, a British expert in environmental risk and disease, who was found dead in his home near Norwich, England, naked from the waist down and wedged under a chair; Tanya Holzmayer, who worked as a microbiologist near San Jose and was shot seven times by a former colleague when she opened the door to a pizza delivery man; David Wynn-Williams, who studied microbes in the Antarctic and was hit by a car while jogging near his home in Cambridge, England; and Steven Mostow, an expert in influenza, who died when the plane he was piloting crashed near Denver.

The stories have also made their way into the e-mail in-boxes of countless microbiologists. Janet Shoemaker, director of public and scientific affairs for the American Society for Microbiology, heard the tales and points out that her organization alone has 41,000 members, meaning that the deaths of 11 worldwide, most of whom were not technically microbiologists at all, is not statistically surprising. "We're saddened by anyone's death," she says. "But this is just a coincidence. In another political climate I don't think anyone would have noticed."

Ken Alibek heard them, too, and dismissed them. Alibek is one of the country's best-known microbiologists. He was the No. 2 man at Biopreparrat (where Victor Pasechnik also worked) before he defected and now works with the U.S. government seeking antidotes for the very weapons he developed. Those who have died, he says, did not really know anything about biological weapons, and if there were a conspiracy to kill scientists with such knowledge, he would be dead. "I considered all this a little artificial, because a number of them couldn't have been considered B.W. experts," he says with a hint of disdain. "I got an e-mail from Pasechnik before he died, and he was working on a field completely different from this. People say to me, 'Ken, you could be a target,' but if you start thinking about this, then your life is over. I'm not saying I'm not worried, but I'm not paying much attention. I'm opening my mail as usual. If I see something suspicious, I know what to do."

Others are not quite as sanguine. Phyllis Della-Latta is the director of clinical microbiology services at New York's Columbia Presbyterian Medical Center. She found an article on the deaths circulating in the most erudite place—an Internet discussion group of directors of clinical microbiology labs around the world. These are the people who, when a patient develops suspicious symptoms, are brought in to rule out things like anthrax.

Della-Latta, whom I know from past medical reporting, forwarded the article to me with a note: "See attached. FYI. Should I be concerned??? I'm off on a business trip to Italy tomorrow & next week. If I don't return, write my obituary."

She now says she doesn't really believe there is any connection between the deaths. "It's probably only coincidence," she says, then adds: "But if we traced back a lot of things that we once dismissed as coincidence—foreigners taking flying lessons—we would have found they weren't coincidence at all. You become paranoid. You have to be."

Don Wiley's body was finally found on Dec. 20, near Vidalia, La., about 300 miles south of where he disappeared.

The Memphis medical examiner, O.C. Smith, concluded that yellow paint marks on Wiley's car suggest that he hit a construction sign on the Hernando DeSoto Bridge, as does the fact that a hubcap was missing from the right front tire. Smith's theory is that heavy truck traffic on the bridge can set off wind gusts and create "roadway bounce," which might have been enough to cause Wiley to lose his balance after getting out of the car to inspect the scrapes. He was 6-foot-3, and the bridge railing would have only come up to mid-thigh.

"If Dr. Wiley were on the curb trying to assess damage to his car, all of these factors may have played a role in his going over the rail," Smith said when he issued his report. Bone fractures found on the body support this theory. Wiley suffered fractures to his neck and spine, and his chest was crushed, injuries that are consistent with Wiley's hitting a support beam before he landed in the water.

The Wiley family considers this case closed. "These kinds of theories are something that's always there," says Wiley's wife, Katrin Valgeirsdottir, who has heard all the rumors. "People who want to believe it will believe it, and there's nothing anyone can say."

The Memphis Police also consider the case closed, and the local office of the F.B.I. has turned its attention to other odd happenings. The talk of Memphis at the moment is the bizarre ambush of the city's coroner last month. He was wrapped in barbed wire and left lying in a stairwell of the medical examiner's building with a live bomb strapped to his chest.

Coincidentally, that coroner, O.C. Smith, was also the coroner who did the much-awaited, somewhat controversial autopsy on Don Wiley.

What are the odds of that?

The numbers 9/11 (9 plus 1 plus 1) equal 11, and American Airlines Flight 11 was the first to hit the twin towers, and there were 92 people on board (9 plus 2), and Sept. 11 is the 254th day of the year (2 plus 5 plus 4).; What of the deaths of nearly a dozen scientists? Is it really possible that they all just happened to die, most in such jarring ways, within so short a time?; The fact that personal attachment adds significance to an event is the reason we tend to react so strongly to the coincidences surrounding Sept. 11. In a deep and lasting way, that tragedy feels as if it happened to us all.; Presidents Kennedy and Lincoln were elected 100 years apart. Both men were succeeded by Johnsons, who were also born 100 years apart. Their names each contain seven letters. Their successors' names each contain 13 letters. Their assassins's names each contain 15 letters.; In a room of 23 people, the odds are even that two of them will have the same birthday. Reduce the occupants to 14, and it's even money that two of them were born within one day of each other. And for a 50-50 chance of finding a pair with birthdays a week apart, you need to invite only seven to the party.

Lisa Belkin is a contributing writer for the magazine.

Study Questions for Case Study 6.4

"The Odds of That" by Lisa Belkin
New York Times Magazine
August 11, 2002

1. What are the two definitions of coincidence given in the article?

2. How does Persi Diaconis describe an unusual day?

3. What is the difference between conspiracy and coincidence?

4. How does Bradley Efron describe how we should measure probability?

5. What does it mean to say that the selection of 20 people is random?

6. What is the probability that at least two people in a random selection of 20 people share a common birthday? In Efron's words, "How surprised should we be if at least two of 20 people share a birthday?"

7. What are the odds against having at least one match among 20 randomly chosen birthdays?

8. What are the odds in favor of having at least one match among 40 randomly chosen birthdays?

9. If the odds against an event E happening are 20 to 1 (20-1), what is the probability that the event E will happen?

10. If the odds against E happening are 7-2, what is the probability of E happening?

11. The article mentions several "unusual" coincidences. Choose two of those and discuss how you view these. Give an example of an unusual coincidence that you or someone you know has experienced.

Additional Exercises and Projects

Additional Exercises for Section 1

1. A new pickup truck costs $23,000 and a new economy sedan costs $14,000. Compare these two costs by filling in the blanks in the following:

   a. _____ is larger than _____.

   b. _____ is _____ less than $23,000.

   c. _____ is _____ percent more than $14,000.

   d. _____ is _____ times $14,000.

   e. $14,000 is _____ percent of $23,000.

2. Estimate the following by performing mental calculations:

   a. $29 \times 301 =$

   b. $28 \div 3.11 =$

   c. 3% of 220 =

   d. 0.5% of 400 =

   e. $53 + 37 + 102 + 98 + 10 =$

3. Find common units for the following pairs of quantities and express the sum of the two quantities in the common units.

   a. 5 feet + 3 yards

   b. $\dfrac{2}{3} + \dfrac{5}{11}$

   c. 6 apples + 8 bananas

   d. $578 million + $4.6 billion

   e. 4 tons + 230 pounds

4. The following are driving distances (in miles) from Los Angeles, CA, to various cities:

San Francisco, CA: 383 miles

Denver, CO: 1019 miles

Kansas City, MO: 1621 miles

New York City: 2785 miles

Orlando, FL: 2516 miles

Las Vegas, NV: 271 miles

a. Compare the distance from Los Angeles to Kansas City to the distance from Los Angeles to San Francisco using a ratio.

b. Compare the distance from Los Angeles to Orlando to the distance from Los Angeles to Denver using a percent.

c. The distance from Los Angeles to New York is what percent of the distance from Los Angeles to Las Vegas?

Projects for Section 1

P1.1 Cost of Iraq War (Case Study 1.1)
The article "What $1.2 Trillion Can Buy" was written while the Iraq War was still being waged. Using information from a reputable Internet site, find a more current estimate for the cost of the Iraq War. Compare this more recent estimate with the author's. Based on the source of your information, would you expect the information you found to be biased in any way? Explain.

P1.2 Understanding a large quantity of money (Case Study 1.1)
Identify a recent large expenditure of money as reported in a media article (e.g., the building of a football stadium, the beautification of a downtown area, or the amount a city spends on public transportation). Write a two to three page paper arguing for alternatives for spending the money. Include reasons as to why your alternatives are preferable.

P1.3 The Dow Jones Industrial Average (Case Study 1.2)
Using information from a reputable internet site, write an explanation of how the Dow Jones Industrial Average (DIJA) is computed. Include some history of the DJIA and a simple example to illustrate the computation. Determine whether Daniel Gross' explanation of the flaws in the DIJA is correct.

P1.4 Unemployment (Case Study 1.2)
Using information from a reputable internet site, write an explanation of how the US unemployment rate is computed. Include a specific scenario (with variations over time) in your explanation. Explain both the strengths and weaknesses of this definition.

P1.5 The Standard and Poor's 500 (Case Study 1.2)
Use your own words to explain (by finding and reading a reputable internet site) how the Standard and Poor's 500 is computed. Include a simple example. Explain both the strengths and weaknesses of the Standard and Poor's 500.

P1.6 Ambiguous Comparisons (Case Study 1.4)
Find five other instances of comparisons such as "times more/times less" or "times higher/times lower" or "percent more/percent less" in advertisements. For each comparison, explain if the comparison is reasonable and comment on whether the comparison follows the strict language interpretation. If not, explain what was probably meant.

P1.7 Is The Number Reasonable?
Find a recent newspaper article that refers to a number that is given with no justification. Determine whether this number is reasonable. Explain your work.

P1.8 "Counting: Use Strawberry Jam," Chapter 1, *The Numbers Game*
Write a two to three page paper that compares and contrasts ideas from the Case Studies in Section 1 with Chapter 1 of *The Numbers Game*.

Additional Exercises for Section 2

1. Express the following tax rates as percents.

   a. 52 mills

   b. $47 per $1000

   c. $29.34 per $326

   d. 48.25 mills

   e. $10,640 on an income of $56,000

   f. $0.83 on a purchase of $9.50

   g. $71,200 on an income of $200,000

   h. $389 on $10,000

2. Fill in the blanks with the missing value.

   a. 75 is _____% of 625.

   b. _____ is 42% of 6700.

   c. 25 is 80% of _____.

   d. 112 is 140% of _____.

   e. 10,000 is _____% of 1500.

   f. 18 is 0.3% of _____.

   g. _____ is 386% of 24,704.

   h. 9 is _____% of 108.

   i. _____ is 0.5% of 400.

3. The number of students in a library increases by 20% after the dinner hour. If 300 students were in the library at the dinner hour, how many were in the library after dinner?

4. If you place $15,000 down on the purchase of a home and this represents 5% of the total price, what is the cost of the home?

5. If the population of New York City was 8,274,527 in 2007, and this represented approximately 42.5% of the population of the entire state of New York, what is the approximate population of New York State?

6. The population of Springfield, Missouri, in 2006 was 150,797, while the population of Columbia, Missouri, in the same year was 94,428.

   a. The population of Springfield is what percent of the population of Columbia?

   b. The population of Columbia is what percent of the population of Springfield?

   c. The population of Springfield is what percent more than the population of Columbia?

7. If $15 socks are on sale for $12, what is the percent reduction (decrease)?

8. If the revenue from a city's sales tax was $570,000 in 2011 and this represented a 3% drop from the previous year's total, what was the revenue from the city's sales tax in 2010?

9. Sasha collected 525 signatures for a petition this week, and this was a 5% increase from the number of signatures she collected the week before. How many signatures did Sasha collect the previous week?

10. The price of milk fell 6% over the past two months to $3.30 a gallon. What was the price of milk two months ago?

11. Suppose that your tuition for this semester is $3500 and that this is a 6% increase over your tuition for last semester. What was your tuition last semester?

12. You purchased gasoline for your car this morning for $2.05 per gallon. This cost per gallon is 12% less than the cost of gasoline one month ago. What was the cost per gallon of gasoline one month ago?

13. The price of a gallon of gas was $3.25 at the beginning of 2008. The price rose 30% by the summer of 2008 but then fell 60% by the end of 2008. What was the price of a gallon of gas at the end of 2008?

14. The stock price for a particular company was $55 at the end of 2008. It fell 40% in 2009 but then rose 15% in 2010. What was the stock price in 2010?

15. In the 2000 census, the population of Bedford was 35,000 and 35% of the population held a college degree. In the 2010 census, Bedford's population had risen to 39,000, and 42% of the population held a college degree.

    a. From 2000 to 2010, what is the increase in the percent of Bedford residents who held a college degree?

    b. From 2000 to 2010, what is the increase in the number of Bedford residents who held a college degree?

    c. From 2000 to 2010, what is the percent increase in the number of Bedford residents who held a college degree?

    d. From 2000 to 2010, what is the percent increase in the percentage of Bedford residents who held a college degree?

16. Suppose a $25,000 car was marked down to $20,000. What was the percent reduction?

17. If the temperature rose from 45°F on Wednesday to 55°F on Thursday, what was the percent increase in the temperature over these two days?

18. A survey was conducted of 2000 people. Of the 2000 people, 1200 were male and 800 were female. Of the males interviewed, 750 said they enjoyed mathematics, while 650 of the females responded that they enjoyed mathematics. (The remainder of the respondents said they did not enjoy mathematics.) Constructing a tree diagram may help you solve the following questions.

   a. What percent of female respondents said they enjoyed mathematics?

   b. What percent of respondents said they enjoyed mathematics?

   c. What percent of respondents who enjoyed mathematics were male?

   d. What percent of male respondents said they enjoyed mathematics?

   e. What percent of female respondents did not enjoy mathematics?

   f. What percent of respondents who did not enjoy mathematics were female?

Projects for Section 2

P2.1 Letters to the Editor on Tax Rates (Case Study 2.1)
Write your own Letter to the Editor that gives a complete and correct explanation of the tax rates from Massery's Letter to the Editor. Include an explanation of errors that appeared in the letters provided in this Case Study.

P2.2 16 to 24 Year Old High School Dropouts (Case Study 2.4)
Use information from Case Study 2.4 and the internet to approximate the percentage of 16 to 24 year olds (male or female) who are high school dropouts. Explain both your reasoning and any assumptions that you made.

P2.3 Find a recent newspaper article that deals with the 2010 Census in your state. Compare the 2010 data with a previous census year. Compute (and explain) at least three absolute changes and three relative changes.

P2.4 Introducing Two Way Tables
There are many useful schematics that allow us to organize data. One organizational method is called a two way table. Below is a two way table that records the votes of members of the US House of Representatives on a bill to implement the United States–Korea Trade Agreement (HR 380, October 12, 2011, www.congress.gov).

	yeas	nays	not voting	Totals
Republican	219	21	1	241
Democratic	59	130	3	192
Totals	278	151	4	433

Use this table to determine the percentage of voting Republicans who voted against the bill and the percentage of voting Democrats who voted against the bill. Use the internet to determine the total number of members of the House of Representatives in October 2011. Then determine the percentage of Republican House Representatives voting in favor of the bill and the percentage of Democratic House Representatives who voted in favor of the bill. Show all calculations and explain your work.

P2.5 "Comparison: Mind the Gap," Chapter 11, *The Numbers Game*
Write a two to three paper that compares and contrasts ideas from the Case Studies in Section 2 with Chapter 11 of *The Numbers Game*.

Additional Exercises for Section 3

1. Answer the following:

   a. Create a set of 8 data points such that the mean is 60 and the median is 65.

   b. Add two numbers to your data set such that the mean increases and the median decreases. What are your new mean and median?

2. Create a set of 10 data points such that all but one of the data points is above average.

3. To receive an A in a particular course, Don needs an average of at least 90 on five exams. His grades on the first four exams were 72, 96, 95, and 89. What is the exact minimum score he needs on the fifth exam to receive an A in the class?

4. Given that the median price of a home in the U.S. was $174,700 on February 20, 2009, determine the cost of living index or home price for the following cities on the same day. (Round all home prices to the nearest dollar, and round all index values to 2 decimal places.)

City	Cost of Living Index	Home Price
Fayetteville, AR	93.59	
Wichita, KS		93,000
Chicago, IL	136.52	
Honolulu, HI		465,000

5. The cost of a barrel of oil was $37.42 in 1980. Use 1980 as a base year and fill in the blank cells in the following table. (Round prices and indexes to two decimal places.)

Year	Cost of barrel of oil	Cost-of-Oil Index
1970	3.39	
1980	37.42	100
2000	27.39	
2005		133.73

   What cost of a barrel of oil would result in a cost-of-oil index of 400?

6. Answer the following:

   a. A person has a BMI of 24 and weighs 54 kilograms. How tall is this person in meters?

   b. Compute the height of a person in feet and inches (rounding to the nearest inch) who has a BMI of 26 and whose weight is 180 pounds.

   c. Using the BMI formula, what would a male who stands 6 feet, 2 inches tall have to weigh in order to have an "acceptable" BMI of 25? (Round to the nearest pound.)

   d. A person has a BMI of 23 and is 1.5 meters tall. What is the person's weight in kilograms?

7. The "Big Three" Internet Stock Index (BTISI) consists of Ebay, Amazon, and Google, and is computed as the S&P 500 is computed. Suppose that the base year was 2002 and the three companies had the following situations in 2002 and 2008.

Company	2002			2008		
	Shares	Price	Capitalization	Shares	Price	Capitalization
Ebay	300	$32		1500	$36	
Amazon	700	$50		900	$65	
Google	1000	$102		1700	$375	

   If the initial BTISI was set at 25 in 2002, fill in the capitalizations and compute the BTISI in 2008. (Round to two decimal places.)

8. The ABC-3 Index is computed as the S&P 500 is computed. The base year is 1995 and the three companies had the following situations in 1995 and 2009.

   a. If the initial ABC-3 was set at 50, fill in the capitalizations and compute the ABC-3 in 2009.

Company	1995			2009		
	Shares	Price	Capitalization	Shares	Price	Capitalization
A	1500	$15		5500	$32	
B	7500	$42		15500	$45	
C	2700	$17		4000	$38	

   b. If the ABC-3 Index is at 180 a year from now, what would the sum of the market capitalizations be for companies A, B, and C?

9. The Midwest Stock Average (MSA) begins with four stocks—A, B, C, and D—and will compute the MSA the way the Dow Jones Industrial Average is computed. At the beginning the stocks sell for $8, $20, $42, and $36 per share respectively.

   a. Compute the initial MSA.

   b. One year later there have been no stock splits and the stocks sell for $12, $19, $52, and $39 when the $52 stock splits 2 for 1 and the new price per share is $26. Find the MSA and the new MSA divisor.

   c. Another year later there have been no additional splits and the stocks sell for $11, $21, $23, and $41. Find the MSA.

10. Three stocks—Ebay, Amazon, and Google—make up the "Big Three" Internet Stock Average (BTISA) that is computed the way the Dow Jones Industrial Average is computed. Assume that no stock splits have occurred since the BTISA was first calculated. Yesterday, Ebay's stock sells for $15, Amazon's stock sells for $66, and Google's stock sells for $348.

   a. Find the BTISA for yesterday.

   b. However, after yesterday's trading, Amazon's stock splits 3 for 1 and the new price per share is $22. Find the new BTISA divisor after the stock split. (Round to three decimal places.)

   c. One month later, there have been no stock splits and Ebay's stock has risen to $22 while Amazon's stock price increased to $38. If the BTISA is 147.85, how much is Google's stock? (Round to the nearest whole dollar amount.)

11. Suppose that gas cost $0.79 a gallon in 1978. Furthermore, assume that because of inflation the cost of gasoline was $3.29 in 2008. Based on the cost of a gallon of gas, what is the value of a 1978-dollar in 2008-dollars?

12. Convert the following units of measure:

   a. 456,192 inches = _____ miles

   b. 8,590,000 cm^2 = _____ m^2

   c. 43,000 mg = _____ kg

   d. 40 ft^3 = _____ in^3

   e. 120 yards = _____ inches

   f. 2 miles^2 = _____ ft^2

   g. 55 miles/hour = _____ ft/sec

   h. 3120 oz = _____ lbs

13. Determine the following:

   a. I am _____ years old. (give as decimal answer, e.g., 19.34 years)

   b. I am _____ days old. (remember to account for leap years)

   c. I am _____ hours old.

   d. I am _____ minutes old.

Projects for Section 3

P3.1 The Gini Coefficient
Use the internet (or print sources) to research the Gini Coefficient. Write a two to three page paper that discusses the history and uses of this statistical measure and determine if the Gini Coefficient is an index. Use some Gini Coefficients to describe and investigate a present day topic of interest to you.

P3.2 Increases in Tuition
Determine the tuition at your college for the past 10 years. Describe how tuition has changed during this 10-year period. Which year(s) saw the greatest change? Using the Consumer Price Index, adjust the tuition figures for inflation. Using the adjusted figures, which year(s) saw the greatest change? Has tuition at your college risen faster than inflation, slower than inflation or kept pace with inflation?

P3.3 Obesity (Case Study 3.4)
Find a recent newspaper article that discusses obesity in the United States. Include several comparisons (both absolute and relative) in your work. You may compare obesity levels among different states and/or obesity levels among different age groups, gender groups or educational groups.

P3.4 NASDAQ Composite Index (Case Study 3.5)
Investigate how stocks are chosen to be included in the NASDAQ Composite stock index and how this index is computed. Write an explanation of why the NASDAQ index is important and whether or not it is more important than the Dow Jones Industrial Average.

P3.5 More on the US Stock market (Case Study 3.5)
Determine what fractions of the total US market (in terms of number of stocks and in terms of market capitalization) the S&P 500, NASDAQ Composite, and DJIA measure. Based on this information write an argument about which of the three is most important as a measure of the health of the US economy.

P3.6 Investing in the Stock Market (Case Study 3.5)
Suppose you have $5000 to invest in the stock market. Determine how you would invest this money. During the next four weeks, track your investments by recording their values on Tuesday and Friday. (This will give you eight specific data points for each stock.) Present your data in an organized manner and comment on your gains (or losses). Describe another investment plan that would have given you a better outcome.

P3.7 Measuring Spending and Revenue (Case Study 3.7)
Update the graphs that appear in Case Study 3.7 by finding data from 2004 through the present. Explain how you found the recent data and how you completed the necessary calculations.

P3.8 Earnings Benefits of College
Choose a profession that normally does not require a four-year college degree, such as welding. Compare the expected income of a welder with that of a high school teacher over a 25-year career. Based on this comparison and other considerations (such as working conditions) argue whether or not a college degree is worth the cost. Of course, you will need to determine a reasonable cost of a college degree.

P3.9 "Averages: The White Rainbow," Chapter 5, *The Numbers Game*
Write a paper that compares and contrasts ideas from the Case Studies in Section 3 with Chapter 5 of *The Numbers Game*.

Additional Exercises for Section 4

1. If you deposit $2,300 in a savings account that pays 3.2% interest annually, find the amount of money in the account after 12 years if the interest is compounded:

   a. Annually.

   b. Quarterly (four times a year).

   c. Monthly.

2. If you deposit $855 in a savings account that pays 4% interest annually, find the amount of money in the account after 5 years if the interest is compounded:

   a. Annually.

   b. Quarterly (four times a year).

   c. Monthly.

3. Instead of buying a car, you decide to invest the money in an account that advertises an annual interest rate of 5.6%. If the interest is compounded annually, about how long do you have to wait for your money to double in value?

4. Suppose you deposit $125 every month in an account that pays 3.5% annual interest compounded monthly. How much money is in the account 2 years after you open it?

5. Amy decides to place $600 in the bank on January 1st each year beginning in 2009 in an account that pays 6.5% interest compounded yearly, with the interest being compounded and added to the balance on December 31 of each year. Use the sum and seq commands on your calculator and compute the amount of money in the account on December 31, 2013. Give the command that you entered into your calculator.

6. Suppose Amy decides to divide the $600 up evenly and place the amount in an account at the beginning of every month, with the account compounding interest monthly on the final day of the month.

   a. What amount is she placing in the bank each month, and what is her monthly interest rate?

   b. Use the sum and seq commands on your calculator to find the amount of money in the account on December 31, 2013. Give the command that you entered into your calculator.

7. Suppose Holly is enrolled in 15 credit hours this semester. The following table displays her grades in the five courses:

Course	Number of Credit Hours	Grade Earned
Math Reasoning	3	A
Geology	3	C
Sociology	3	A
Chemistry	4	B
History	2	B

Calculate Holly's GPA for the semester (round to two decimal places).

8. A student is taking 12 hours in her final semester and needs at least a 3.5 GPA in this semester to graduate with honors. She is taking a 3-hour Math Reasoning course where she assumes she'll make an A, and a 3-hour English course where she'll make a B.

   a. If she has a 4-hour biology course and a 2-hour lab course where her final grade is unknown (assuming she already knows her Math and English grades), how many remaining grade points does she need to earn in the biology and lab courses to finish with a 3.5 GPA for the semester?

   b. Using this answer, what are all the ways she can finish the semester with at least a 3.5 GPA?

9. Assume that you are considering two options when buying a new car. A gasoline version of a Honda Civic costs $16,000 and gets 38 MPG, while the hybrid version of the same car costs $24,000 and gets 62 MPG. Assume that you drive 18,000 miles a year and that gasoline costs $2.85 per gallon.

   a. Find the cost equation for both the gasoline version of the Honda Civic (C_g) and for the hybrid version (C_h).

   b. Sketch the graph of the two equations.

   c. If you plan to own the car for 10 years, which car would make the most sense to buy? Why? How many years will it take to break even on the extra cost of buying the hybrid?

10. Assume the yearly savings in gasoline costs due to the purchase of a Hybrid automobile are $850 and the extra cost of purchasing the Hybrid was $5500.

   a. Produce an equation of net savings over a period of x years.

   b. Graph this equation and identify the point that tells the number of years required to break even. What is the number of years needed to break even?

11. Suppose that the difference in price between a gasoline vehicle that gets 35 MPG and the hybrid version of that same vehicle that gets 50 MPG is $2448. Suppose also that a car dealer tells you that if you drive on average 14,000 miles a year, it will take you **exactly** 6 years to pay off the difference in buying the hybrid over the gas version. If he is correct, what was the car dealer's estimate for the average price of a gallon of gas over these 6 years?

Projects for Section 4

P4.1 Analyze a Credit Card Statement (Case Study 4.1)

Find a current credit card offer and create a one year scenario for yourself. You must make at least three purchases every month. For the purposes of this project, do not pay more than 50% of your monthly bill. Calculate the monthly finance payments and track your credit card balance over this year.

P4.2 Saving for a Purchase (Case Study 4.2)

Choose an object that you want to purchase some number of years after you graduate from college. For example, you may want to purchase a $40,000 boat ten years after you graduate. Write a paper analyzing at least two reasonable savings plans that will result in your ability to obtain the $40,000 in ten years.

P4.3 Saving Money by Doing Without (Case Study 4.2)

In Case Study 4.2, the author discusses the amount of money accumulated by a person who gives up a daily cappuccino, saves $2.50 a day for 25 years, and invests the proceeds at 8% interest. Consider something you regularly spend money on (e.g., a yearly expenditure for season tickets to a sports team, a monthly expense of a pedicure, or daily eating out at a restaurant). What if you gave this expense up and instead invested the money? Identify an available savings account that pays a particular interest rate and repeat the scenario given in Case Study 4.2 based on your sacrifice and the interest rate on your account. How much savings would your sacrifice produce after 25 years?

P4.4 Present Day Comparison of Car Models (Case Study 4.3)

Update Case Study 4.3 by comparing a car that comes in both a hybrid and a traditional gas model. Determine reasonable values for the sales price and mileage per gallon (mpg) rating for each model. Personalize this analysis to your own situation by determining how many miles you normally drive during a year and by using a gasoline price for your location. Explicitly state all assumptions that you make as you determine how long you will need to drive and keep the car in order to "break even." Project Extension: Extend this analysis to include putting the money that you save by buying the traditional model into a savings account according to current interest rates for your location.

Additional Exercises for Section 5

1. The following table gives the number of homes listed for sale in four cities in two consecutive months.

City	March 2009	April 2009
Smallville	345	415
Midville	605	665
Central City	720	700
Westend	450	405

a. Construct a bar graph to represent the number of homes listed in the four cities in March 2009.

b. Construct a line graph to represent the number of homes listed in the four cities in March 2009.

c. Which of the two graphs from (a) and (b) best represents the data for March 2009? Explain your reasoning.

d. Find the percent changes in the number of homes from March to April in the four cities and graph these changes using a bar graph.

2. Suppose that 34% of all college students in the United States attend private colleges, 25% of all the college students in Michigan attend private colleges, and 41% of all college students in Kentucky attend private colleges. Give reasons why these data are not correctly represented by the pie chart below.

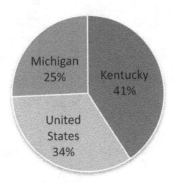

3. Assume that the monthly percent changes in the cost of a gallon of gasoline are given in the following table, beginning January 1, 2009, and ending January 1, 2010.

Jan 1 to Feb 1	Feb 1 to Mar 1	Mar 1 to Apr 1	Apr 1 to May 1	May 1 to Jun 1	Jun 1 to Jul 1	Jul 1 to Aug 1	Aug 1 to Sep 1	Sep 1 to Oct 1	Oct 1 to Nov 1	Nov 1 to Dec 1	Dec 1 to Jan 1
10%	5%	−5%	−8%	15%	20%	15%	8%	−5%	5%	10%	8%

   a. Construct a bar graph that represents these 12 percent changes.

   b. Assuming that gasoline costs $2.00 a gallon on January 1, 2009, compute the costs of a gallon of gasoline on the first of the next twelve months, ending on January 1, 2010.

   c. Construct a bar graph of the costs you computed in part (b).

4. The percent change in the cost of flying during the given years is presented in the following graph. Suppose it cost on average $250 to fly at the beginning of 2005. Give the expected costs for the beginning of each year in a table and produce a bar graph of these costs on the given axis.

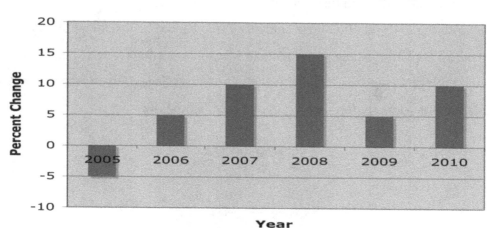

Percent Change in the Cost of Flying

Year	2005	2006	2007	2008	2009	2010	2011
Cost							
Change over past year	xxx						

Projects for Section 5

P 5.1 Enrollment at your University (Case Study 5.1)

Examine the enrollment figures for your university over the past ten years. Create a graphic similar to the ones in Case Study 5.1, and use your data to determine what the average rate of growth was over the past ten years. Use this average growth rate to project enrollment figures for the next five years.

P5.2 More on Tax Cuts (Case Study 5.3)

Create an Excel spreadsheet that determines the "pre tax cut" and the "post tax cut" amounts paid by each income group in Case Study 5.3. (You will have to come up with a plan to deal with the fact that one graph includes eight income groups and the other includes nine income groups. Explain how you chose to handle this.) Create a stacked column chart that shows both the post tax cut amount for and the amount saved by each income group.

P5.3 Examining More Recent Tax Cuts (Case Study 5.3)

Find data on past tax cuts (e.g., the 2001 Bush tax cut, the 2003 Bush tax cut, or others) and determine how the tax cut was spread across income groups. Write a two to three page paper discussing your findings and the implications stemming from the chosen tax cut.

P5.4 U.S. Health Care Spending (Case Study 5.5)

Use the internet to obtain the 2010 national health care spending amount. Refer to Case Study 5.5 and create a new graph of "The Costs of Health Care, and the Shifts in Coverage" and use reliable sources to predict the 2020 amount.

P5.5 "Up and Down: A Man and His Dog," Chapter 4, *The Numbers Game*

Write a paper that compares and contrasts ideas from the Case Studies in Section 5 with Chapter 4 of *The Numbers Game*.

Additional Exercises—Section 6

1. In how many different ways can you choose a clothing combination if you have 6 shirts, 4 pairs of jeans, and 3 pairs of shoes from which to choose?

2. Suppose that a restaurant menu lists 9 different appetizers, 18 options for the main course, and 7 options for dessert. In how many different ways can you choose an appetizer, main course, and a dessert from the menu?

3. Suppose you have five dice, each one a different color. In how many different ways could the outcomes on the dice appear if you rolled all five dice at the same time?

4. Evaluate the following:

   a. $5!$

   b. P_7^{10}

   c. C_{10}^{14}

   d. $\dfrac{9!}{7!}$

5. A committee of 5 is to be formed from a group consisting of 12 women and 7 men.

   a. How many different committees are possible?

   b. How many of the possible committees have exactly 3 women as members?

   c. How many of the committees have at least one man as a member?

6. There are 9 different positions on a baseball team. In how many ways could these positions be filled from a group of 15 people?

7. There are 5 girls and 4 boys heading to a movie. If each person purchases his or her own ticket, in how many ways could they stand in line if we consider only the gender of the person standing in line? (EX: GGGGGBBBB is one way they could stand in line)

8. A company sends you a rewards card that allows you to choose 3 magazines to receive free from a list of 20 publications. In how many ways could you choose your 3 selections?

9. A committee of three is to be formed from a set of people consisting of three Democrats, five Republicans, and two Independents.

   a. How many different committees can be formed (regardless of party affiliation)?

   b. How many committees would consist of all Republicans?

   c. How many committees would consist of no Independents?

   d. What is the probability of choosing a committee at random that did not have any Republicans?

10. Assume there are 366 days per year.

    a. How many different possibilities are there for the 8 birthdays of 8 people?

    b. How many different possibilities are there for the 8 people under the condition that no two people share the same birthday?

    c. What is the probability that no two people in the group of 8 share the same birthday?

    d. What are the odds against no two people in the group of 8 sharing a birthday?

11. Assume that a meteorologist states that there is a 30% of rain today.

    a. What is the probability it will not rain today?

    b. What are the odds against it raining today?

12. One student is to be selected at random from a group of 8 freshmen, 5 sophomores, 8 juniors, and 4 seniors.

    a. What is the probability that the student is either a freshman or sophomore?

    b. What is the probability that the chosen student is not a freshman?

    c. What are the odds against the chosen student being a freshman?

    d. What are the odds against the chosen student being either a freshman or a sophomore?

13. A student flips a coin 6 times.

    a. What is the probability the coin lands on heads (H) all 6 times?

    b. What are the odds against the coin landing tails (T) all 6 times?

    c. What is the probability that 3 flips will results in heads (H) and 3 will result in tails (T)?

    d. What are the odds in favor of getting exactly 4 heads?

14. Assume that the following are odds against winning for each of six horses in a race.

Tommy Lee	3-1	Crazy Legs	9-1
Gravel Road	7-2	Run Around	4-1
Tim Tam	7-1	Lightning	9-1

    a. What are the odds against the winner being Tommy Lee, Crazy Legs, or Lightning?

    b. What are the odds in favor of the winner being Run Around or Tim Tam?

Projects for Section 6

P6.1 Taking Lipitor (Case Study 6.1)

Research the clinical advantages and disadvantages of Lipitor. Do your findings agree with the statistics reported in the ad from Case Study 6.1? Discuss at least two advantages and two disadvantages of taking Lipitor.

P6.2 Understanding Medical Two Way Tables (Case Study 6.2)

Measuring a man's prostate-specific antigen (PSA) is one method that has been used to screen men for prostate cancer. Sometimes a PSA value above 4.0 is used to indicate the possible presence of cancer. If someone does not have cancer, their PSA value will be 4.0 or smaller 94% of the time. However, if someone does have cancer their PSA value will be above 4.0 only 20% of the time. This information can be better understood when organized in a two-way table.

a. Consider a study which involves 1000 men, 900 of whom do not have prostate cancer. If each individual in this study is screened for cancer using the PSA, the following table illustrates the likely results:

	Has cancer	Does not have cancer	**Total**
PSA value above 4.0	20	54	**74**
PSA value 4.0 or below	80	846	**926**
Total	**100**	**900**	**1000**

You should be able to check that 94% of the men who do not have cancer had a PSA value of 4.0 or below while 20% of those with cancer had a PSA value above 4.0. The above table can also be used to answer additional questions regarding the accuracy of the PSA test. For instance, of all the men who had a PSA value above 4.0 (i.e., they tested positive for cancer), what percentage did not have cancer? This is called the *False Positive Rate.* Of all the men who tested negative for cancer (PSA value of 4.0 or below), what percentage actually had cancer? This is called the *False Negative Rate.*

b. Suppose you duplicated this experiment with another group of 1000 men. However, this time assume that only 10 of the men actually have cancer. Create the new two-way table and calculate the false positive and false negative rates.

P6.3 State Lottery (Case Study 6.4)

Find a recent news article on your state lottery (or, if your state does not have a lottery, choose a nearby state which does). Determine the lottery income, individual winnings and the state profit for a specific period of time. Discuss any controversies that may have occurred.

P6.4 Kentucky Derby (Case Study 6.4)

Find a news article that details the odds on at least ten horses from a recent Kentucky Derby. Compute the probability of a win for each of the ten horses. Discuss the race results and explain how much people won, based on specific bets.

P6.5 "Chance: The Tiger that Isn't," Chapter 3, *The Numbers Game*

Write a two to three page paper that compares and contrasts ideas from the Case Studies in Section 6 with Chapter 3 of *The Numbers Game*.

Selected Answers to Additional Exercises

Selected Answers to Additional Exercises for Section 1

1. A new pickup truck costs $23,000, and a new economy sedan costs $14,000. Compare these two costs by filling in the blanks in the following:

   a. $23,000 is larger than $14,000.

   b. $14,000 is $9,000 less than $23,000.

   c. $23,000 is 64.3 percent more than $14,000.

   d. $23,000 is 1.643 times $14,000.

   e. $14,000 is 60.9 percent of $23,000.

3. Find common units for the following pairs of quantities and express the sum of the two quantities in the common units.

   a. 5 feet + 3 yards = 5 feet + 9 feet = 14 feet, or 1.67 yards + 3 yards = 4.67 yards

   b. $\dfrac{2}{3} + \dfrac{5}{11} = \dfrac{22}{33} + \dfrac{15}{33} = \dfrac{37}{33}$

   c. 6 apples + 8 bananas = 6 pieces of fruit + 8 pieces of fruit = 14 pieces of fruit

   d. $578 million + $4.6 billion = $578 million + $4600 million = $5178 million, or $.578 billion + $4.6 billion = $5.178 billion

   e. 4 tons + 230 pounds = 8000 pounds + 230 pounds = 8230 pounds, or 4 tons + .115 pounds = 4.115 tons

Selected Answers to Additional Exercises for Section 2

1. a. 5.2%   b. 4.7%   c. 9%   d. 4.825%   e. 19%

   f. 8.74%   g. 35.6%   h. 3.89%

3. 360

5. 19,469,475

7. 20%

9. 500

11. $3301.89

13. $1.69

15. a. 7 percentage points   b. 4130 people   c. 33.7%   d. 20%

17. 22%

Selected Answers to Additional Exercises for Section 3

1. Many possible solutions, for example:

    a. 40,40,40,64,66,76,76,78

    b. 40,40,40,<u>59,64</u>,64,66,76,76,78/new mean is 64.3 and new median is 64.

3. 98

5. To get a Cost-of-Oil Index equal to 400, a barrel of oil would have to cost $149.68

Year	Cost of barrel of oil	Cost-of-Oil Index
1970	3.39	**9.06**
1980	37.42	100
2000	27.39	**73.20**
2005	**50.04**	133.73

7.

	2002			2008		
Company	Shares	Price	Capitalization	Shares	Price	Capitalization
Ebay	300	$32	9600	1500	$36	54000
Amazon	700	$50	35000	900	$65	58500
Google	1000	$102	102000	1700	$375	637500
Total			146600			750000

Capitalization in 2002 was 146,600. Capitalization in 2008 was 750,000. BTISI in 2008 was 127.90. Calculation: $(750000/146600)(25) = 127.899$.

9. a. Initial MSA is $(8 + 20 + 42 + 36)/4 = 26.50$.

    b. New MSA is $(12 + 19 + 52 + 39)/4 = 30.5$ and the new divisor is $x = (12 + 19 + 26 + 30)/30.5 = 3.15$.

    c. New MSA is $(11 + 21 + 23 + 41)/3.15 = 30.48$.

11. $1 in 1978 dollars is equivalent to $4.16 in 2008 dollars. (Set up $.79/1 = 3.29/x$ and solve for x.)

13. Varies per person

Selected Answers to Additional Exercises for Section 4

1. a. $2300(1.032)^{12} = \$3356.48$

    b. $2300(1 + .032/4)^{48} = \$3371.58$

    c. $2300(1 + .032/12)^{144} = \$3375.01$

3. After 12 years, a \$10,000 investment will yield \$19,229 while after 13 years it yields \$20,306. So an investment will double in about 13 years. Using logarithms: $\ln(2)/\ln(1.056) = 12.7211$, so a more precise answer would be 12.7 years.

5. $600(1.065^5-1)/.065 = \$3416.18$, subtract 600 since we're asking for Dec. 31.

    $3416.81 - 600 = 2816.81$

    $\text{sum}(\text{seq}(600*1.065^x,x,1,4,1)) = 2816.18$

7. $(3(4) + 3(2) + 3(4) + 4(3) + 2(3))/15 = 3.20$

9. a. Cost of gas per year for the gas model is $(18000)(2.85)/38 = 1350$

    Cost of gas per year for the hybrid model is $(18000)(2.85)/62 = 827.42$

    $Cg(x) = 1350x + 16,000$ $Ch(x) = 827.42x + 24,000$

    b.

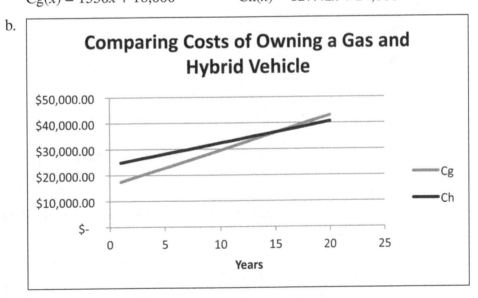

    c. If you only keep the car for 10 years, the gas version makes more sense because on the interval $[0,10]$ $Cg(x)$ is less than $Ch(x)$. In fact, it would take about 15.3 years to break even.

11. $(14000/35)x - (14000/50)x = (2448/6)$. Solve for x to get \$3.40 per gallon of gas.

Answers to Additional Exercises for Section 5

1. a.

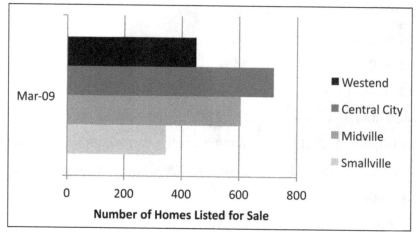

b.

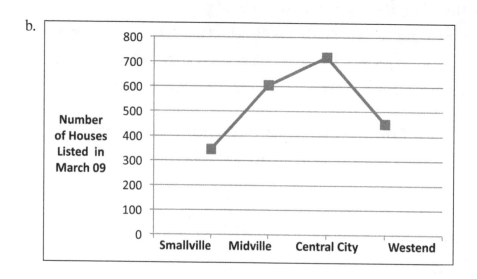

c. The graph in (a) is better, because there is no reason to connect the dots in graph (b).

d.

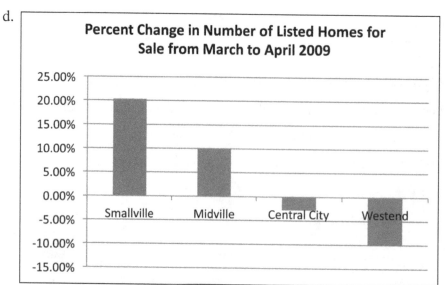

3. a.

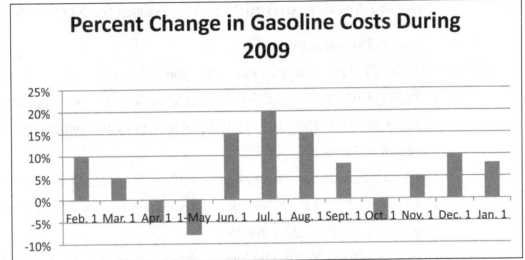

b.

Jan. 1	Feb. 1	Mar. 1	Apr. 1	May 1	Jun. 1	Jul. 1	Aug. 1	Sept. 1	Oct. 1	Nov. 1	Dec. 1
$2.00	$2.20	$2.31	$2.19	$2.02	$2.32	$2.79	$3.20	$3.46	$3.29	$3.45	$3.80

c.

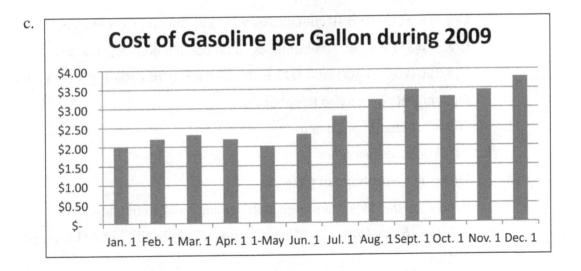

Selected Answers to Additional Exercises for Section 6

1. 6(4)(3) = 72. 72 combinations.

3. 6^5 = 7776. 7776 possible outcomes when you roll 5 dice.

5. a. C(19,5) = 19!/(14!5!) = 11628 possible committees of 5 from the 19 people.

   b. C(12,3)C(7,2) = (12!/(9!3!))(7!/(5!2!)) = 4620 possible committees containing 3 women and 2 men.

   c. Determine the number of committees that have 5 women. This will be C(12,5) = 12!/(7!5!) = 792. So the number of committees with at least one man will be 11628 − 792 = 10836.

7. C(9,5) = 9!/(4!5!) = 126 different line formations, considering only the gender of the 9 people.

9. a. C(10,3) = 10!/(7!3!) = 120 is the number of ways to create a committee of 3 from a group of 10 people.

   b. C(5,3) = 5!/(3!2!) = 10 committees would contain no Republicans.

   c. C(8,3) = 8!/(5!3!) = 56 committees could contain no independents.

   d. C(5,3)/C(10,3) = 10/120 = 1/12 is the probability of choosing a committee at random that contains no Republicans.

11. a. 70% or .7 chance of no rain

    b. 7 to 3 odds against rain

13. a. 1/64 is the probability of 6 Heads

    b. 63 to 1 are the odds against the coin landing Tails all 6 times.

    c. C(6,3) = (6)(5)(4)/6 = 20, 20/64 = 5/16 is the probability of 3 Heads and 3 Tails

    e. C(6,4) = (6)(5)/2 = 15, odds in favor of getting exactly 4 heads are 15 to 49.

Additional Readings to Augment the Course

Better: A Surgeon's Note on Performance, by Atul Gawande, Picador, 2008.

Bringing Down the House: The Inside Story of 6 MIT Students who Took Vegas for Millions, by Ben Mezrich, Free Press, 2003.

Math for Life: Crucial Ideas you Didn't Learn in School, by Jeffery Bennett, Roberts and Company Publishers, 2012.

Mathematics and Democracy—The Case for Quantitative Literacy, edited by Lynn Arthur Steen, NCED, 2001. Available online at http://www.maa.org/ql/mathanddemocracy.html

Moneyball, by Michael Lewis, Free Press, 2003.

Quantitative Literacy—Why Numeracy Matters for Schools and Colleges, edited by Bernard L. Madison and Lynn Arthur Steen, NCED, 2003. Available online at http://www.maa.org/ql/qltoc.html

Stat-Spotting: A Field Guide to Identifying Dubious Data, by Joel Best, University of California Press, 2008.

The Numbers Game, by Michael Blastland and Andrew Dilnot, Gotham Books, 2009. This book was published in the United Kingdom under the title, *The Tiger That Isn't.*

Why Numbers Count—Quantitative Literacy for Tomorrow's America, The College Board, edited by Lynn Arthur Steen, 1997.